ETHNOGRAPHIES OF CONSERVATION

ETHNOGRAPHIES OF CONSERVATION

Environmentalism and the Distribution of Privilege

Edited by David G. Anderson and Eeva Berglund

Berghahn Books
New York • Oxford

First published in hardback in 2003 by

Berghahn Books

www.BerghahnBooks.com

First paperback edition published in 2004

Library of Congress Cataloging-in-Publication Data

Ethnographies of conservation : environmentalism and the distribution of privilege /
edited by David G. Anderson and Eeva Berglund.
  p. cm.
Includes bibliographical references and index.
ISBN 1-57181-464-7 (cloth : alk. paper) ; ISBN 1-57181-696-8 (pbk: alk. paper)
  1. Human ecology--Moral and ethical aspects. 2. Human ecology--Economic
aspects. 3. Consumption (Economics)--Moral and ethical aspects. 4. Environmental
ethics. 5. Distribution (Economic theory)--Moral and ethical aspects. 6. Distributive
justice. 7. Material accountability. 8. Conservation of natural resources--Economic
aspects. I. Anderson, David G., 1965- II. Berglund, Eeva K.

GF80 .E855 2002
178—dc21

2002066724

British Library Cataloguing in Publication Data

A catalogue record for this book is available from the British Library

Printed in the USA on acid-free paper.

ISBN 1–57181–464–7 hardback
ISBN 1-57181-696-8 paperback

Figure 1 Replant, recycle, repair the world? The grass airstrip at the heart of the conservation area in Haia, Papua New Guinea. *Photo*: David M. Ellis.

Contents

MAPS AND FIGURES

ACKNOWLEDGEMENTS

Editing this collection was significantly helped by the fact that, as Dawn
Chatty suggested, the chapters were submitted for a round of comments
among the contributors themselves. We would like to thank Sean Kingston
at Berghahn Books for his support throughout the production of this volume. We would also like to thank Dr Mike Taylor, now working in
Botswana, for the excellent work he put into drafting the maps and for
helping us compile the manuscript. We are very grateful to Carole Berger
for her help with the proofs and index.

We gratefully acknowledge the support of the Department of Anthropology at Goldsmiths College for their help in hosting the original seminar
from which this collection arose.

NOTES ON THE CONTRIBUTORS

Cristina Adams is a Ph.D. candidate at the Department of Ecology, University of São Paulo, Brazil. She was Visiting Research Fellow at the Department of Anthropology, University of Kent at Canterbury, 1999–2000. Her research experience concerns environmental anthropology, protected areas, the Atlantic rainforest and the Amazon.

David G. Anderson is senior lecturer in social anthropology at the University of Aberdeen. His ongoing research is on ecology and indigenous rights in the circumpolar Arctic. He has recently published a monograph on Evenki and Dolgan reindeer herders entitled *Identity and Ecology in Arctic Siberia: The Number One Reindeer Brigade* with Oxford University Press.

Eeva Berglund was lecturer in anthropology at Goldsmiths College. She has written on the anthropology and history of environmental politics. Her monograph *Knowing Nature, Knowing Science: An Ethnography of Environmental Activism* was published in 1998.

Dawn Chatty is Senior Dulverton Fellow and Deputy Director of the Refugee Studies Centre, Queen Elizabeth House, University of Oxford. Her research interests include pastoral nomadism and conservation, gender and development, and coping strategies of children and adolescents in prolonged conflict and forced migration. Her monograph *Mobile Pastoralists: Development Planning and Social Change in Oman* was published in 1996.

David Ellis is a Ph.D. student in environmental anthropology at the University of Kent at Canterbury and works for the Future of Rainforest Peoples Programme. He conducted fieldwork in Papua New Guinea between 1996 and 1999 on the comparative ethnography of local histories, and the impacts of conservation biology and development. He is the author of 'Pawaia people of the Pio-Tura region, Papua New Guinea': Final Report of the Future of Rainforest Peoples Programme to the European Commission.

Nicola Frost is a Ph.D. student in the Anthropology Department at Goldsmiths College, University of London, with an ongoing research interest in politics and development in Eastern Indonesia.

Bornali Halder is completing her D.Phil. at the Institute of Social and Cultural Anthropology at the University of Oxford. Her main research interests are Native North American philosophies, indigenous rights, environmental activism and environmental ethics.

Dario Novellino is a Visiting Research Associate of the Institute of Philippine Culture (Ateneo de Manila University), and a Ph.D. candidate in environmental anthropology at the University of Kent at Canterbury. His current research interests include indigenous metaphysics, the techno-symbolic value of tool-behaviour and technological choices, ethnobiology, indigenous rights and environmental conservation.

Stephen Nugent teaches in the Department of Anthropology, Goldsmiths College, University of London and is Associate Fellow of the Institute of Latin American Studies. He is the author of *Big Mouth: The Amazon Speaks* (1990) and *Amazonian Caboclo Society: An Essay on Invisibility and Peasant Economy* (1993). With Mark Harris, he is currently preparing an edited volume, *Some Other Amazonians*.

Anja Nygren is an environmental anthropologist at the University of Helsinki, Finland. She is currently visiting scholar at the School of Natural Resources, University of Missouri-Columbia, United States. Her current research interests include environmental conflicts, protected areas, cultural representation of tropical forest-dwellers, legal pluralism and diversity of local livelihood strategies. She is the author of *Forest, Power and Development: Costa Rican Peasants in the Changing Environment*.

Luna Rolle holds a Master's degree in environmental anthropology from the University of Canterbury at Kent. She currently works with the Association ATEMA assisting the socio-economic integration of the Gypsy population in Strasbourg. Her current interests are in youth education and intercultural learning in interethnic contexts.

Sian Sullivan is a British Academy Post-Doctoral Research Fellow in the Deptartment of Anthropology at the School of Oriental and African Studies, London University. Her academic interests include cultural landscapes, dryland ecology and resource use, 'community'-based conservation, environment and development discourses, gender, dance and 'the body'. Among her recent publications is *Political Ecology: Science, Myth and Power* (co-edited with P. Stott).

Rachel Wrangham is a social anthropologist who has conducted research in Indonesia since 1995. She is currently working on her Ph.D. at the London School of Economics. Her current work focuses on rural development in Mozambique, and she has also worked in India and Thailand. She is an Associate Professional Officer with the United Kingdom Department for International Development.

INTRODUCTION
Towards an Ethnography of Ecological Underprivilege

Eeva Berglund and David G. Anderson

Anthropologists at the start of the twenty-first century face familiar professional challenges. As always, they must engage with both the ethnographic encounter and the preoccupations of academic production. Once again, as a century ago, anthropological work is framed by the crumbling of relatively stable global power blocks. If a century ago the so-called centres of world civilisation found themselves dethroned by colonial independence movements, and then gutted by world wars, today a new order is being built upon the rubble of post-socialism and warring fundamentalisms. At the start of the past century, the great settler states of Argentina, Australia, Brazil, Canada, Russia and the United States launched continental ecological experiments, transforming grassland into farmland and forest into plantations. At the start of this century, intensifying competition to control strategic minerals and fuels continues to parcel out local ecosystems with pipelines, railways, and transitory settlements, and, more likely than not, it is altering the planet's climate. If in the nineteenth century botanists prospected among indigenous societies for useful plants, today biotechnologists race to license the genetic patterns of flora and fauna. Everywhere they go, anthropologists learn of the struggles that accompany these contests, yet their reflections on these encounters remain marginal to public and political debate. This book arises out of a sense that the discipline could and should reshape itself through an active engagement with these debates, alert to the echoes of the not dissimilar debates which shaped the discipline a century ago.

Seen in historical relief, the identity of the discipline appears fundamentally challenged by the new politics of ecology. At the start of the twentieth century, anthropologists were not yet 'professionals'. Nonetheless, when reading colonial blue books or even missionaries' accounts, we can still recognise ourselves in the actions of those who were concerned about the plight of rural peoples in colonial hinterlands. In particular, we like to claim a kinship with those 'practical men' who spoke against slavery and structured exploitation, and who drafted enlightenment programmes for the 'protection' of languages, cultures and traditional lands (Kuklick 1991). On this topic, however, we find an important contrast with the applied work of our twenty-first-century professional colleagues. The anthropological community, hardened by decades of struggle for recognition in various academies, is no longer confident that it is worthwhile to classify and collect folklore 'before it disappears', let alone contribute to 'the betterment of savages'. Most anthropologists today would also be uncomfortable with the idea of engaging with 'traditional' communities. And so it is that many sit at the sidelines while not-yet professional environmentalists fight public battles to protect what they confidently perceive as untouched landscapes inhabited by traditional peoples. Does our absence from these battles stem from a wisdom already won through having travelled a similar path? Or does it have more to do with the possibility that professional anthropologists of the twenty-first century have lost faith in their ability to transform the arena of action out there into apposite knowledge, and that we are simply more comfortable occupying ourselves with texts and cultural critique?

Looking back on ourselves one hundred years ago we now puzzle at the heady mixture of imperial politics which lent a heavy hand in the drafting of ethnographies. As George Stocking (1991) noted, the ethnographer of the pristine and the exotic arrives on the scene only a decade or two after mercenary armies have pacified the local population. How is the ethnographic enterprise timed today? Have we been reduced to the role of somewhat naïve chroniclers of local diversity supplying picturesque material to colour in the outlines of serious ecological operations? Or, might it still be that ethnographic observation could help us get our bearings in this new order of conservation, consumption and new information technologies?

The chapters in this book were each drafted to answer questions such as these. The thirteen scholars printed here, each writing an ethnography of ecological politics in different parts of the world, were participants in a one-day workshop held at Goldsmith's College, University of London, in April 2000. All but one of the participants were anthropologists and all have considerable experience in writing ethnographies with an ecological dimension. Meeting under the banner of writing 'Ethnographies of Ecological Underprivilege', we found ourselves sharing, often for the first time, unvoiced thoughts that had arisen while sometimes sitting politely as observers at environmentalist meetings, as consultants charged with writing background pieces for environmental non-governmental organisations (ENGOs), or even as active organisers of campaigns to protect special places in the areas where we conduct fieldwork. In trading our stories, we discovered that though employed in somewhat unfamiliar endeavours, we had nevertheless amassed a fair body of observations, questions and often written notes, which were ethnographic in character but somehow did not quite fit the genre of anthropological ethnography. We also identified a troubling convergence of theme, irrespective of whether we were speaking of the former Soviet Union, the United States of America, Papua New Guinea, Namibia, Syria, or Brazil. We were disturbed to find that there is in this century, once again, a happy fit between expressing discourses of primitiveness and historic relationships of disempowerment – only not within colonial offices but within the cafés and meeting rooms where environmental consultants meet. In many cases the slogans of pristine environments and unspoiled places served to deflect attention from the fact that large-scale circuits of capital perpetuate both ecological and economic degradation. The clues to these interests could often be read directly from the environmentalist discourse, but more often than not, the most interesting issues came from comparing the official transcripts of environmentalist literature with concrete observations of these activists at work in their biotic and social environments. The settings of our ethnographies were not only the often rural locations affected by environmentalist intervention, but the urban offices and park headquarters which themselves thrived upon a complex ecology of government grants, publicity and consistent stereotypes of people and nature 'out there'.

The goal of this volume is to show how an ethnography of environmentalist practice helps to reveal a broader ecology of relationships, such as the fact that ecopolitics involves the redistribution of resources and costs. The chapters document both efforts to address ecological degradation *and* the unequal distribution of environmental hazards. As a collection, they sometimes draw uncomfortable connections between abuses of specific human rights on the one hand, and demands for global ecological health on the other. In this introduction therefore, we take to task the assumptions of virtue embedded in environmentalism, not through a cultural critique of the 'West', but by drawing on the ethnography of environmentally inspired action.

The book is organised into three parts which loosely follow the structure of this introductory chapter. In the first section, ethnographic analyses demonstrate the irreducibly political and distributive dimensions of environmentalist thought and action. In each of the three cases – in Nicaragua (Anja Nygren), Brazil (Cristina Adams) and Indonesia (Nicola Frost and Rachel Wrangham) – the struggle over conservation discourses is inextricably linked to the rights of the people who inhabit the affected regions. The chapters make visible the people marginalised by the 'anti-political' nature of many conservationist accounts.

In the second part, the authors examine how projects to protect spaces are linked to myths of state identity or of national progress and so legitimate differential accumulation of wealth. We learn how conservation invokes metaphors of bounded space to separate people from their lands, but also to discriminate people from people. Bornali Halder and Dawn Chatty examine how, in the United States and in Syria respectively, the state imperative to protect specific spaces for the national good can be used strategically to dispossess aboriginal peoples. Sian Sullivan's account of resistance to conservationist demarcations takes place in Namibia, where the political climate is quite explicitly geared towards nation building. Her empirically rich account clearly demonstrates the complexities and dilemmas involved as globally legitimated conservation projects are translated into everyday practise. David Ellis suggests that conservation in Papua New Guinea actually creates privileged enclaves for consumption at the same time restricting local people's engagements with their environment. Further, his chapter forcefully questions the assumptions of virtue and higher justice embedded in environmentalist rhetoric (Figure 1). All four chapters unfold under the spectre, acknowledged or not, of ethnic discrimination. In this arena anthropology can provide rich insights, valuable precisely because the architects of global conservation would appear to ignore this dimension as troublesome detail, even as their actions are likely to intensify the significance of ethnically based identity politics.

The final section explores the metaphors used in ecological discourses. The first two chapters, by David Anderson and Luna Rolle, analyse the appeal to national myths in post-socialist Russia. In his examination of attempts to restructure hunting economies in Siberia, Anderson shows how urban industrial interests switch metaphoric codes to justify exploitation in a free-market idiom. Rolle's detailed account of young environmentalists in Tatarstan shows how metaphors of nature can capture and promote simultaneously innovative and conservative social and political messages. Supported further by Nygren's study of post-Sandinista Nicaragua (in the first section), these chapters are a reminder that theorising ecodiscourses from an overly Anglo-American perspective risks undervaluing the potential for lessons to be learned from ethnographic but historically informed analysis. The plea for paying greater attention to the lessons of ethnography is most forcefully articulated by Dario Novellino, who compares the understanding of the forest among Philippine aboriginal peoples and that of the environmental legislators who are trying to act on the forest's behalf. He suggests that ecopolitics needs to open itself up to more radical experiments in how we conceive nature. In the volume's concluding chapter by Stephen Nugent, we are given a rich account of how ecological metaphor has structured the history and development of an entire region. This chapter underscores the value of examining the intellectual and not just the moral sources of the muted role of social science in these issues. He gives us occasion to reflect on the positive role that ethnographers can offer the study of environments and ecopolitical action at the start of this century.

In the remainder of this introductory chapter, we shall examine the benefits of ethnographic exploration, suggesting that together with an understanding of the historical emergence of powerful political agendas like environmentalism, ethnography can go far beyond efforts to provide local detail. Anthropological work on conservation is hardly limited to the examples in this volume, but we feel there is work to be done in rendering it more confident. Indeed, the transformation of such work into theory has been surprisingly slow, partly because the object of analysis is politically problematic for anthropology. To generate a robust anthropological contribution to ecopolitics, we propose that what needs to be made visible is not only the effects of environmentalism on marginal peoples in a context of economic polarisation, but the lifeworlds of environmentalists themselves, whether they be part of grassroots campaigns or of international environmental non-governmental organisations.

Anthropology, ecopolitics and discrimination

Although anthropology has developed a fine canon of discourse analysis, the pioneers in the analysis of environmentalist discourse have been environmental historians and cultural theorists. The primary insight of a generation of work on this subject is that environmentalist statements, which are pitched at an angle which avoids discussing people, are nevertheless primarily discourses about political power. If so-called mainstream environmentalism speaks 'against people' on behalf of nature, critics point out that what is defined as natural varies immensely from one historical and geographical context to another (Guha 1990, Cronon 1995).

For example, the rhetoric which fuels hyperbole about the power of genes takes for granted the idea that nature is itself, or contains within it, a force for self-organisation, self-preservation and even self-replication which brings about evolutionary change. Yet for intellectuals as recently as the nineteenth century (as for contemporary Creationists) nature by itself without the Creator's power was/is not thought to have such capacities (Bowler 1993). Nature for gene enthusiasts is hardly the same as it is for Creationists, but nonetheless it is made sacred and essentialised for both. Any effort to save 'it' then is linked to the political question of who should manage it. In the examples provided here, nature becomes the province of experts regardless of who occupies it and, furthermore, provides grounds for discriminating against the very people who do.

For most of our contributors one of the first arenas where environmentalism and discrimination came together was in the creation of protected areas. Indeed fencing off, or otherwise designating places for limited uses, is literally a form of discrimination. Whether these sites are of special scientific interest or royal hunting grounds, the parcelling of the environment differentiates between those with authorised access and those who are excluded.

The first step in discriminating a space is to strip it of human history – to naturalise it. A familiar example is the image of Brazilian rainforests, discussed in this volume by Nugent and Adams, who emphasise that these special spaces have been treated as if they had no politico-economic significance, even though their histories are replete with extreme and expansive violence. They are treated as if they were practically uninhabited except by fauna and by quasi-natural populations of humans. The Amazon in particular is iconic of nature as something sacred and inaccessible. If the image of dense jungle is central to this quintessential nature, so is the idea of the ecologically noble savage. Thus, the Amazonian and Atlantic rainforests provide key examples of how nature, but also primitiveness can be wielded politically, making it easy for any interest – corporate, administrative, but also scholarly (Adams this volume) – to ignore already existing, complex social, ecological and political systems. Although Brazil provides the classic example, this collection shows that locations as varied as Siberia, the Philippines and Papua New Guinea are just as easily made into naïve spaces with local populations, if they are visible at all, expected to either exercise survival skills in perfect harmony with their environment, or to let their homes become biodiversity reserves.

The second step in discrimination is to link spaces to a state myth which legitimates protectionist action. Again, following the lead of our colleagues in environmental history (Nash 1982, Thomas 1983, Harrison 1992), we find that ideas of pristine nature and the moral imperative for conservation are linked to bourgeois ideologies regarding the place of people in society. The classic demonstration of this argument comes from the cultural history of North American environmentalism. It is revealing that the myth of the frontier, which enabled white colonisers to justify the dispossession and slaughter of indigenous populations in the United States, is also the founding myth of American environmentalism (Nash 1982, Cronon et al. 1992). The governing ideologies behind the parks movement were so powerful that most people accepted that the landscapes chosen for protection were in fact natural, pristine and free from human impact. Moreover, the action of creating protected spaces fit easily into familiar models that environments must be appropriated, transformed and altered in order for them to be valuable.

The opening section of the volume clearly illustrates the close links between interventions in the name of conservation and overtly value-laden programmes for the management of peripheral or marginal populations. All three cases are linked to something usefully thought of as global ecopolitics, but they also insist that the precise forms and effects of these links are matters for concrete investigation. In contrast, much influential social science scholarship starts from the lofty axiom that the 'global environmental crisis' is yet another manifestation of the epistemological chaos unleashed as Euro-American modernity implodes (Giddens 1991, Beck et al. 1994, Latour 1998). Some, still haunted by neo-Malthusian fears of the poor and the backward, consider that a planet-wide institution for protecting nature is the only thing that will prevent catastrophe (Meadows et al. 1992, Sachs 1993). As an empirical truth, the increasing density of global communications networks and circuits of resource exchange is taken to mean that the 'global' is the level from which to start analysis. Yet upon closer investigation the ecological risks which are most obviously inscribed into 'global' institutional arrangements, such as Agenda 21 adopted by the United Nations, are to an alarming extent rooted solidly in elite economic and political preoccupations. Susan George (1998: xiii) has made the simple but powerful observation that the way by which ecopoliticians approach the environment makes peripheral people increasingly invisible while photogenic mega-fauna are brought increasingly into focus. The most powerful factor which seems to preselect the objects of nature protection, whether they be flood defences for the English countryside or improved 'status' for big-game animals, seems to be the imperative of capital accumulation. As Stephen Nugent (2000: 241) observes, environmental risk is in fact 'not risk: it is business as usual'.

Ethnographic work can do much to demystify naïve yet politically powerful myth-making. However, as anthropologists we must come to terms with the way that our polite silence in political and professional forums might tacitly help this process. In fact we suggest that the subtle way that resolutions concerning power relations work their way into environmental discourse is aided by a certain reticence to take environmentalists to task. Anthropologists are indeed faced with an ethical dilemma when it comes to conducting 'ecocritiques' (Milton 1996, Luke 1997, Brosius 1999). On the one hand, perhaps out of a sense of professional competition, we feel compelled to introduce nuance into the descriptions of environments as 'pristine', 'untouched' and of local people's ideas being reduced to the minor (technical) role of 'stewards' or 'managers'. On the other hand, when we turn our ethnographic craft to writing about the culture of environmentalists (or other protest groups) we risk violating our own ethical creeds by exposing debates or discussions which our new informants would prefer to keep hidden (Brosius 1999). Often one feels that one is attacking the wrong target. After all, wealthy as they may be, we assume that high-profile or transnational environmental organisations will never shape tomorrow's world as powerfully as the alliances of states, corporations and militaries. However, as Sian Sullivan indicates in her ethnography of a protest aimed at an environmental non-governmental organisation (ENGO) in Namibia, in certain sectors non-governmental organisations can wield power disproportionate to their numbers, such that they can confidently threaten local stakeholders and even scholars who offer polite but pointed criticisms of their actions.

One way of respectfully breaking our silence is to try to understand the ecology of the organisations themselves, their internal relationships and the constraints under which they operate. We can question why and how certain forms of biotic but also social differentiation are invoked in setting the parameters of ecodiscourse (cf. Wilmsen 1996). Thus by applying the same principle that anthropologists have applied to support environmental historians, we can provide an equally valuable service by portraying environmental organisations not as 'naïve organisations' but as relationships shot through with history and politics.

In applying this humanistic perspective, it is easier to recognise that many organisations also have a remarkable capacity for learning, and they respond rapidly to changing political environments. Many major ENGOs have sought ways to incorporate social agendas and nuanced, rather than global, analyses into their actions, albeit with varying degrees of success. However, even where community participation is encouraged ecopolitics nevertheless may have negative political and economic consequences for marginal people. Sullivan, Chatty and Nygren (this volume) show that even in complex and shifting political landscapes, ethnographers can take a stance when ecopolitics ends up legitimating current structures of power. This should, and can be done in a way that opens up a space for dialogue rather than threatening ecopoliticians with censure (cf. Wenzel 1991, Sillitoe 1996, Freeman et al. 1998, Trigger 1998, Robbins 2000).

It is time to provide some provisional definitions. There is a risk that environmentalism may be taken as synonymous with environmentalist discourse. This tendency is amplified by a large literature which seeks to derive the environmentalist interest from a set of core values associated with Western religious traditions, history and economic practices (e.g., Worster 1977, Pepper 1986, Harvey 1993). Here, we are making much more humble claims by striving to make sure that environmentalism has faces, names and biographies. That is, environmentalism, as we use it, refers to a set of practices which flow out of the everyday life of concrete, committed people, many of whom live in the metropoles and not in the hinterlands which they strive to protect.

The term ecopolitics, in contrast, aims to be both broader and narrower in scope. With its prefix 'eco', it draws attention most obviously to both ecological and economic issues. It invokes ideas about the politicisation of physical space (Kuehls 1996), but it also generates a conceptual space (Ó Tuathail and Dalby 1998) where ecological and economic agendas are counterpoised using the image of a specific protected area as a battleground. In that sense, where ecopolitics is championed by people with protectionist ideologies, it encompasses environmentalism. In another sense, however, ecopolitics is a narrower term than environmentalism. For unlike environmentalism, ecopolitics refers to something that is always already politicised in an explicit sense. Whether in the World Bank's Global Environmental Facility or in a regional sustainable transport campaign, ecopolitics highlights the question: in whose interest are problems defined as urgent? We would like to see this politicised and conceptual use of the term ecopolitics aligned with our programme of writing ethnographies of ecological practice. After all, the original root of the word ecology – oekos – also denotes the household as much as it denotes economy (Goodland 1975). We see it as important that our work specifies for whom economic and ecological issues are dearest and who fights for them from the home-front.

In sorting through these issues on the political meaning of ecology, the insights of anthropologists are useful. Anthropological analysis has suggested that the environment is more than 'all that surrounds' (Ingold 2000: Ch. 12). It is instead a sphere of life activity – a place where one dwells and makes a life for oneself. The chapters in this book, each of which examines in considerable detail the interplay between power, economics and environmental projects, all take as a given that any park or reserve cannot be a moral good in its own terms, but can only be judged by its success at enabling an integration of human and nonhuman life within a specified region. Taken in this sense, understandings of environmental discrimination, even if necessarily power-laden, can be applied constructively in talking about the world – but only if they include a range of interests wider than those of environmental organisations themselves.

Distributing justice within protected landscapes

Environmental historians have alerted us to the fact that meditations on space can lead into discourses on political liberty and larger discussions of how to define the 'good life'. Recent historical research work in the 'mirror-empire' to the United States has suggested new ways of thinking about conservation and parks. In light of conventional wisdom about the need for global environmental protection regimes, the idea of state socialist or Soviet environmental movements seems like an oxymoron. However, environmental historians have discovered an interesting contrast to the textbook example of bourgeois nature protectionism in the example of socialist views on the relation between society and nature (Weiner 1988, 1999; Pyne 1997). In Douglas Weiner's work we learn of a vibrant ecological movement in late Imperial and early Soviet times which fought a spirited but unsuccessful battle against the industrial 'promethianism' which came to mark Soviet socialism after the 1930s. The quality of this movement is intriguing for it did not focus upon the contemplation of nature in solely an aesthetic sense, but saw the need for nature (*zapovedniki*) as places where biologists and ecologists could protect baseline examples of 'pure' nature before it was transformed by human forces. Early Russian studies, based in these so-called scientific preserves, anticipated what is now regarded as modern thinking about trophic energy transfers and neo-Lamarckian models of adaptation by thirty years. Similarly, many of the main actors who led this programme of scientific protection also led the movement to create autonomous spaces for indigenous people where their economies and cultures could be developed independently along state socialist lines.

These views of directed development have their own ideological element of underprivilege. As most accounts of socialist approaches to nature and to territory tell us, collectivisation economists rode roughshod over local models of how best to relate to nature. However, unlike North American models in which individual emancipation is discovered through aesthetic contemplation on the empty spaces of the frontier, the state socialist models show how the good life could be imagined by studying the ways human agency could be applied to nature. In the words of Douglas Weiner (1988), the nature reserves of the former Soviet Union could be understood as 'little islands of freedom' wherein critical discussion on the future could be camouflaged as scientific measurement of biological process.

The habit of imagining the good life through imposing a territorial form upon nature has not been completely silenced in the former Soviet Union. In fact it permeates today's debates and it is at this juncture that ethnography can make an important contribution. As Rolle argues in her highly detailed discourse analysis of the Guard of Nature movement in Tatarstan, its environmental activists see themselves as the vanguards of democratic opposition against the Soviet past. The statements reproduced in her account clearly illustrate the power that the idea of nature has for bracketing discussions on human rights. The history of the region's environmental activism is beyond the scope of Rolle's chapter, but it is significant that Tatarstan established one of the first locally run reserves of the former Soviet Union in 1926 (Weiner 1988: 256). Anderson examines the struggles for survival of hunters and reindeer herders in central Siberia. He relates the story of the rapid and 'successful' disengagement of the Soviet state from any interest whatsoever in the well-being of its rural citizens, citing as its reason the need to return native people to the forest. Left on their own in forests depleted of game and without the most basic inputs needed for hunting, Evenki and Yakut people search for an avenue to advertise their shocking new disempowerment enforced on purely ecological grounds. Their proposals, like so many in the state socialist tradition, show a combination of economically astute ideas for local exchange networks embedded within radical proposals for economically engaged parks and reserves. Here, as in Tatarstan, parks become a metaphor for the discussion of human rights.

Centralised nature management is not only an artefact of Eastern redistributive states. Nygren's analysis (in Part I) of local struggles over nature protection in post-Sandinista Nicaragua shows a similar authorised model of development embedded within an agricultural agency which wields authority for developing peoples and lands. The liberal market economies of the West also provide a fertile environment for redistributing rights to land. If we turn back to the classic examples of protected spaces, North America, numerous scholars have given us rich examples of how supposedly wild places have been managed by people and thus have brought the myth-making process into greater relief. The work of the anthropologist Henry Lewis (1989, 1992) and the environmental historian Stephen Pyne (1997) have demonstrated the various ways that aboriginal peoples have used a simple but effective technique – burning – to 'sweep' the landscape of unwanted and often dangerous tinder. Fire technology, however, was immediately identified by state authorities as the antithesis of good management for the simple fact that in their view it destroyed forest cover in order to regenerate it. Although this management style is muted compared to the boundaries, uniforms and signage common in most national parks, it is extremely effective. In fact, it seems that one of the best ways to change a landscape is to deprive it of human activity. In recent research in Canadian national parks, Eric Higgs (2000) has re-photographed late nineteenth century romantic panoramas to demonstrate that a century of management, suppression of fire and the eviction of mixed-blood farmers has made entire valleys into overgrown caricatures of their former 'pristine beauty'.

The distributive effects of centralised nature protection are not, unfortunately, merely a nineteenth-century curiosity. Indeed, as the essays in Part II show, there is evidence that in the past two decades such ideologies of protection have gained a new lease on life. Chatty recounts how Bedouin herders have been displaced from their lands on the assumption that it was their unregulated land-use practices which were spoiling the landscape. Halder shows the systematic ecological violence perpetrated against the Lakota Indians by the state, which has curtailed access to resources and caused environmental destruction through mining and military use. Whereas the state operates on exclusive boundaries when it suits it, the Lakota organisations refuse the flattening out of all space by criss-crossing it with ever more boundaries. Sullivan documents the increasing control of foreign-based NGOs over indigenous lands which builds on the conviction that spaces and species are more important than the local ecology of human relationships. The Batak of the Philippines, described by Novellino in the book's third section, dwell in a world which is also not divided into exclusively owned spaces. Traced against this ethnographic background, the colonising model of a world with absolute boundaries appears as a hardened tool of a conquering power whose authority must be clearly established or not at all.

Disempowerment can also have more traditional economic and symbolic dimensions. The chapters by Ellis on Papua New Guinea and Anderson on Siberia both focus on how ecologically inspired alterations in consumption patterns often work against the economic interests of local people and threaten the local ecology. By building on what anthropologists have frequently observed as a disjuncture between what people say they are doing and what they actually do, Ellis sketches an all-too-familiar picture of the unsustainable consumption habits set up by ENGO workers in the rainforest of Papua New Guinea. As in Anderson's chapter, such practices by conservationists and administrators are often one of the best clues to the true meaning of environmentally inspired action. It is at such junctures that the critique of statement and of texts relies upon politically evocative ethnography. By writing an ethnography of ecological underprivilege, anthropologists can go beyond simple environmental or historical critique and can help forge a new dialogue on the justice of how people and space are related.

Writing environmentalism

In his survey of engagements between anthropologists and environmental movements, Peter Brosius (1999: 277) argues, 'If ever there was a rich site of cultural production, it is in the domain of contemporary environmentalism'. All of the chapters here bear out the spirit of this statement. They concur on the need to look at how nature is represented and how people like those we study get caught up and implicate themselves in these processes. As Novellino stressed during the workshop, and as Nugent argues in the volume's concluding chapter, there is a currency of symbols and language which circulates through ecopolitics. The reader first feels the power of words in the disorienting effect created by the strings of environmental acronyms. Once one learns to negotiate the landscape of eco-speak, one must learn to wield the key terms of conservation, protection and sustainability. What then are the links between such languages and symbols, and the circuits of material goods and investment, which provide the conditions for societies to be reproduced? These are questions that go to the heart of what it means to write an evocative ethnography of ecological relationships.

Each of the volume's chapters takes a unique position on the value of *writing* as a tool for understanding the ecology of environmentalists themselves, suggesting overall that ethnography can help untangle myth-making operations from the unconscious assumptions that guide ecopolitical action. Drawing on a kind of ethnography which has a commitment to long-term exploration of social relations at various scales, each chapter attends to the cultural and ecological context before turning to an analysis of environmentalist discourse. The ethnographic method, we argue, quite simply helps make more transparent the assumptions of ecopolitics. Yet it is difficult for anthropologists to be too righteous about exposing metaphoric operations within environmentalism. After all, modern anthropology grew out of a similar fusion of moral conviction and an engagement with exotic places as does contemporary environmentalism. We would like to think that having come to terms with some of the contradictions in the worldview underlying this moral view, anthropological fieldwork nevertheless offers important nuances. This is demonstrated in the way each chapter here makes use of ethnography's ability to trace metaphors which aptly yet almost imperceptibly capture sets of power relationships.

Nygren's analysis of the environmentalist debate in Nicaragua encourages us to pay attention to the struggles over 'environmental images'. In Río San Juan the way that the jungle was described by different actors either as 'wild' or as 'productive' gives important clues to the pattern of power in a regional setting. Ironic in this case is the fact that both state managers and local *campesinos* described themselves as engaged in a struggle with nature. However, the state managers characteristically saw themselves as the much more virile and powerful actors with a paternal responsibility to act over and above 'their' people on the ground. In the chapter by Frost and Wrangham, the Indonesian forest provides a wealth of symbols which competing political actors struggle to use. In this context, it is those who are able to distribute their ideas most widely, over the Internet or locally in campaign literature, who can win the authority to speak for the forest. The struggle to earn the right to speak about a special place takes on a litigational accent in the chapter by Sullivan. Here ecopolitics is so intensely focussed on authority, that people and places become but faint shadows obscured by ecological concepts. We are reminded of some of the metaphoric limits of ecopolitical discourse in Halder's chapter on the struggle of the Lakota Sioux for control over their site of spiritual renewal – the Black Hills of the north-eastern United States. Here the struggle takes place directly in court – an arena of words and interpretation. Through legal transcripts we learn that Lakota Sioux refuse to accept monetary recompense for damages that are spiritual. Instead, they would prefer to legislate new metaphors for their landscape, ones which would evoke the depth of seven generations of kinsfolk, rather than the shallow values of commodifiable natural resources.

One of the important conclusions that our analyses of metaphor and method have reinforced is that ENGOs are complex and heterogeneous, and one should not assume that the organisational form of an environmental movement, be it an ENGO, part of a donor agency, or a government institution, will determine the out-

come of an intervention in any sense. From an ethnographic perspective organisational form fades into the background. Rather, what is highlighted is the complex
way individuals and groups are positioned in matrices which are neither 'progressive' nor 'reactionary' (Fisher 1997, Morris-Suzuki 2000). The analysis of metaphor
helps to identify political ideas embedded in ecological agendas, and thus paves the
way towards evaluating them. Through studying the way that people describe their
ecology and their economy, ethnography allows us to appreciate Dawn Chatty's
irony that 'conservation promotes poaching' for Bedouin herders. David Ellis points
out more mundanely how 'conservation is about consumption' in the Papua New
Guinean rainforest. Anderson and Novellino highlight the various uses of the words
'market' and 'globe' as clues to more complex ecological relationships than those
set by politicians. The political question remains, of course, whether those who lose
out in ecopolitics-as-usual can make their own practises count so that they might
flourish and generate more hopeful practices. An unexpectedly positive example is
the case presented by Anderson where the thoroughly corrupt application of the
market metaphor in Siberia can nevertheless be shown to create certain new spaces
wherein locally meaningful action can be reproduced.

As we noted, the paradoxes which ethnographic analysis brings out are not at
all new to social theory. Both the words environment and nature overflow with
metaphoric power, and both have been shown to be politically and intellectually
troublesome, to put it mildly. Nature's ontological primacy has been shaken in
social theory (Strathern 1992, Latour 1993, Descola and Pálsson 1996a, Braun and
Castree 1998) as well as in ethnographic accounts (Strathern 1980, Descola 1993,
Tsing 1993, Scott 1996, Descola and Pálsson 1996b, Ellen and Fukui 1996). These
analyses have shown that both European and North American common-sense definitions of nature posit a sharp division between the manipulating, subjective
human and passive, objective nature. Today, however, the distinction between
technology as human innovation and nature as pre-cultural seems to have
imploded in practice as well as theory. For example, machines now help people
live out their 'natural' capacity to be parents (Strathern 1992: 177), plants are valued for their information content – genes – and even more strangely, 'nature' now
requires culture in the form of environmentalism simply to continue to exist.

In the practice of conservation but also in technology policy, the nature–culture dichotomy still carries politically crucial moral messages (Hayden 1998,
Franklin et al. 2000). One key forum where dualistic metaphors of nature and culture run rampant is in the discussion of global versus local perspectives. As global
ecopolitics so often works to the detriment of those whom anthropologists study,
our reaction has been to place greater emphasis on local histories and context in
our writing. If, however, anthropology confines itself to describing local events, it
is in danger of falling into impotent particularism at best, and politically suspect
populism at worst. What is needed instead, is an analysis of how local events evoke
political narratives of global scope. In this way we can identify the dynamics which
promote the ever-intensifying circulation of the goods and the bads of which
global ecopolitics is composed. Above all, anthropologists should draw attention
to the use of the concept 'global' in a depoliticised or shorthand way such that it
is coterminous with a cold concept of 'nature'. After all, in one sense the Batak

described by Novellino too have a global view, but one where connections between different contexts are not made by reference to a nature taken as isomorphic with the 'globe' (cf. Strathern 1995), rather, they are created through the capacities of beings, human and other, for co-apprehension or mutual recognition. As Novellino demonstrates, thick description can add layers of meaning to concepts which at first glance seem very simple. Nugent gives us another example through his analysis of the slogan that 'nature is out of control'. In his view, such language creates an idiom of quasi-scientific ecologism through which anthropologists too have sought an authoritative voice. Such an idiom may be more tasteful than one of 'civilising the savages', but it nevertheless reproduces a metaphor that people deemed to be 'closer to nature', be they primitives or criminals, must be managed (Luke 1997).

Another danger to which ethnographers can fall victim is to be placed as assessors of 'authenticity' of rural cultures. During the seminar, it came as a surprise to us all how common and pervasive was the idea that only those ecologically 'noble savages' living in a morally strong 'untouched' landscape are entitled to some sort of political support from conservation agencies. We were not so much surprised by the idea itself, which is not new, but by the fact that it is still strong and vital at the start of the twenty-first century. What was also uncomfortable was that many of us felt tempted to encourage this stereotype in the texts of our unpublished reports and 'discussion papers' in order to achieve what we felt to be noble short-term goals of protecting a special place or supporting a movement (cf. Kirsch 2001).

Yet we recognise that the language of authenticity has effects far beyond scholars and activists. Nygren (this volume) reports how difficult it was to persuade local people that an ethnographer was indeed interested in the lives of *campesinos* dressed in polyester.

In thinking about this challenge to sincere ethnography, it is important to recognise that this process is not accidental. The badge of authenticity which comes from protecting a proper natural place or traditional people has been an essential tool to legitimating the services of an ENGO to sponsors (Conklin 1997, Brosius 1999, Anderson and Ikeya 2001). Where more detailed accounts tarnish the prelapsarian credentials of rural people, often environmentalists cease to be interested, continuing to operate instead with stereotyped images of simple peasants or victimised primitives which obscure not only cultural heterogeneity but important political detail. A famous and illuminating example of such simplification are India's Chipko or 'tree-hugging' protesters, made up of heterogeneous constituencies who were nevertheless constructed by admiring Western audiences simply as 'an exemplar of grassroots environmentalism in the Third World' (Rangan 1996: 217). Tragically however, although it was easy for tree-hugging peasants to gain outside support, here as elsewhere, continuing struggles against economic and political marginalisation scarcely register unless the sound of guns can grab the media's attention. There are lessons here about how easy it is to invert politically convenient stereotypes, as yesterday's ecologically noble savages are suddenly turned into murderous ethnic fundamentalists.

Even if it is rare for any one anthropologist to witness such dramatic transformations as they unfold, the power of turns of phrase should never be lost upon us as writers, and may yet yield important insights into the momentum of global ecopolitics. Ellis' account in this volume is by far the most sobering account of how environmental discourse and practice has become so rationalised that it seems to serve the material interest of its leaders rather than engaging with the specifics of the local ecology. In his radical view, talking about the environment is a weak replacement for living in the environment. However, balancing Ellis' account is Anderson's hopeful attempt to highlight the possibilities for new syncretisms between local values and new free-market and environmentalist ideologies.

Having surveyed some of the pitfalls of ecological discourse, and the deliberate or tangential ways that anthropologists participate in it, is it possible to identify a clear path forward? The seminar convinced us that one way forward would be to do what anthropologists have always done best – to provide detailed and thoughtful of accounts of the ways people talk about the world. Unlike our forebearers of the nineteenth and early twentieth centuries, today we recognise the need to analyse not only places far away, but also, and increasingly importantly, cultural milieux like boardrooms, ENGO offices and airport lounges closer to 'home'. Ethnographic thickness and the method's openness to dialogue can render the strange more understandable, but it also helps question, to make strange what is apparently familiar. In a sense, we already know what is needed for such a project: more fieldwork (albeit in different places), an apt lexicon, and a critical eye towards 'new' cultural forms (such as nature 'preserve' or the cosmology of Traditional Ecological Knowledge). Yet we also worry that this might be an easy and self-aggrandizing option. After all, the call for such a renewed ethnographic project carries the implicit message that we must have more grants, more studentships and more publications – maybe even new departments! Does such a productivist philosophy bring us closer to providing influential ethnographies? The answer might lie in anthropology's past.

We argued above that anthropology became estranged from the local ecology of power and action through its professionalisation. This suggests that we should also pay more attention not only to how we write but to the genres we produce. As the examples in this collection show, ethnographies of ecological underprivilege should be reflexive, sensitive to local turns of phrase, and also canny. However, it seems that they should also take calculated risks. All the authors in this volume have made choices about how to name the places they have lived and the people with whom they worked. In qualifying some of their 'data' they implicitly show that their texts engage with real, important and often difficult human dilemmas. Secondly, during the seminar all of the authors in this collection expressed dissatisfaction both with the formal public transcripts that we were taught to write as students and the grey, unpublished texts that we are often contracted to write as collaborators in ecological projects. The texts in this volume include material which is normally not included in the formal, refereed genre of anthropological ethnography. By stretching the genre we hope to give more exposure to these silent texts, as well as give expression to those critical thoughts that we have been led to hide. The reverse project, which is beyond this volume, would be to inject the skill of ethnographic precept and critique into development projects and its 'grey' literature, perhaps making it more reflexive and hopefully making it more public. Many of the chapters here allude to attempts, successful and unsuccessful, at negotiating such a path.

Perhaps one of the best ways of illustrating our vision of writing ethnographies of ecological underprivilege is to report the challenges which arose in editing this volume. We hope that the reader will agree that the material is rich and diverse, but as such it posed familiar problems in terms of legislating consistency in spelling and style. As is traditional, we as editors take full responsibility for these changes. Some of our contributors ran up against the usual shortcomings of this academic genre in that ethnographic sections had to be edited for length and reworked to make them accessible to readers who may not be expert in each of the four continents considered in this volume (see Figure 2). This we found to be a serious problem, but we hope it is not fatal since there will, we hope, be other volumes and other journal articles by each contributor on the topics dealt with here.

However, we were less well prepared for a challenge that touches directly on the issues we set out to analyse. We were drawn into a pointed electronic mail exchange with an organisation which questioned not only points of fact but also the line of inquiry that the author in question was following. The correspondence even hinted at litigation. In trying to mediate the dispute we discovered that changing the pages in question was not the issue. Instead, the problem appeared to be that our efforts had moved the anthropological gaze towards relatively powerful organisations without giving these organisations the right of veto. Obviously anthropology is always to some extent an intervention that requires the negotiation of awkward relationships of trust, responsibility and power at all stages (Brosius 1999, Kirsch 2001). In unpacking this incident we found out that like other metaphors analysed here, the word 'responsibility' evokes a complex world of fact, interpretation and certain cultural intangibles such as reputation. We have to report that we are not such heroes as to have forged ahead in reproducing the pages in question unchanged. They were subject to no less than ten rounds of editing. We are confident that they represent no prejudice to the persons, places or organisations named herein. However, this caveat does not provide the same security as that offered by complete silence. The fact that these pages appear in print at all is in its own way ethnographic evidence of the significance that this line of inquiry might hold. Indeed, the entire volume is an exploration of the possible impacts of shifting the research focus, and of the surprises ethnography might have in store.

The chapters in this volume all attest, we feel, to the transformative potential of ethnographic work. The influence of political ecology as an academic as well as politically engaged field of enquiry is palpable in this volume, as is the legacy of a post-structuralist sensibility to discourse. Drawing on the best tradition of critical social science, the chapters shed light on hitherto overlooked aspects of environmentalism, making full use of ethnographic objectification. The outcome of all social activity, including anthropological writing as well as environmentalism, is of course contingent and so our objectifications will never remain within our control, nor will they ever, of course, be anything but partial (in both senses). This means that in writing of ecology and conservation as ways of distributing privilege, we should be prepared to enter dialogue with, not just write about, the actors in our ethnographies.

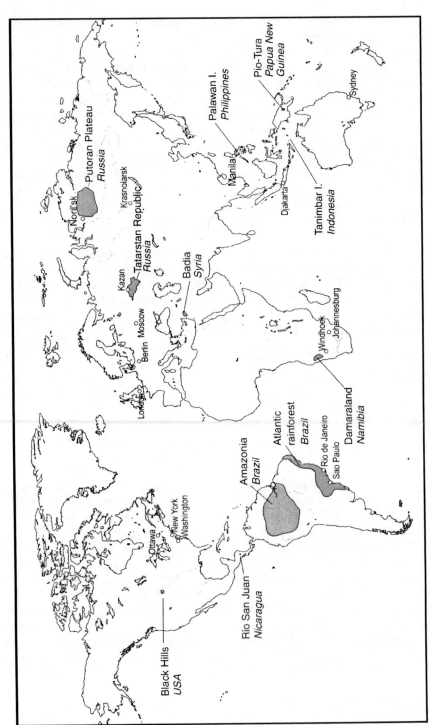

Figure 2 World map

PART I

ANTHROPOLOGY, ECOPOLITICS AND DISCRIMINATION

1

PITFALLS OF SYNCHRONICITY

A Case Study of the *Caiçaras* in the Atlantic Rainforest of South-eastern Brazil

Cristina Adams

> To successfully resist ongoing systems of domination, racial or ethnic stereotyping, and cultural hegemony, the first necessity of disempowered peoples, or of marginalized subcultural groups within a national society, is that of constructing a shared understanding of the historical past that enables them to understand their present conditions as a result of their own ways of making history. (Hill 1996: 17)

Introduction

The first protected area to be established in Brazil was the Itatiaia National Park, in 1937, inspired by the North American conservation model. Since then many Protected Areas have been instituted by the Brazilian government, and divided into categories of restrictions. According to national environmental laws,[1] restrictive protected areas (national parks, biological reserves, ecological and biological stations) should be used only for scientific, leisure or conservation purposes, remaining uninhabited. As in most developing countries, the establishment of protected areas in Brazil was a top-down process motivated more by political than by technical reasons. A significant number of protected areas were defined on a map after a flight over the region.[2] As a result, many restrictive protected areas were established in places already occupied by human populations, especially indigenous people. According to law, these peoples should have been expropriated, receiving payment in exchange for their land. In reality, in most cases nothing was done about it, with the result that populations found themselves illegally inhabiting their own ancestral lands, prevented from practising subsistence activities, such as swidden cultivation, hunting, collecting, fishing and extracting wood, fibres, fruits and nuts (Vianna et. al.

1995). Similar conservation strategies have produced the same outcome in the Philippines, Nicaragua and the United States (Novellino, Nygren and Halder respectively, this volume).

Encouraged by international discourse, the debate over whether or not to remove human populations from restrictive protected areas in Brazil gained momentum from the 1980s. Many sections of national society were involved: government agencies, non-governmental organisations (mostly of environmentalists, Roman Catholic or Pentecostal churches, local inhabitants, politicians) as well as the academic community. As elsewhere (Ellen 1993, Milton 1993), debate became polarised between two groups: 'anthropocentrists' on the one hand, championing the permanent right of indigenous populations to occupy their ancestral homes, and 'conservationists' supporting their removal on the other. The former believe that indigenous or traditional subsistence activities are based upon ancient awareness of the habitat and do not affect ecosystems in a negative way. Indigenous people would be an example of humanity living in a 'harmonious' relationship with nature. The latter base the argument for removal on the fragility of ecosystems and maintain that local people cause degradation.

This chapter is concerned with the particular debate regarding the protection of the last remnants of the coastal Atlantic Rainforest growing in the south and south-eastern coast of Brazil, inhabited by indigenous *Caiçara* populations.[3]

When the Portuguese conquerors first contacted the Amerindians in 1500, the Atlantic Rainforest covered most of the coast, stretching for more than 4,000 km north to south, from 5° S (Cabo de São Roque, RN) to 30° S (Rio Taquari, RS). After five centuries of exploitation it has been reduced to only 5 percent of its original area (Brown and Brown 1992), and receives less attention from international media and environmental organisations than the Amazon. Figure 3 shows the states of São Paulo, Santa Catarina, Paraná and Rio de Janeiro and existing remnants of Atlantic Rainforest.[4]

The existence of *Caiçara* settlements within protected areas in the Atlantic Rainforest presents a legal problem that should be solved by the resettlement of populations. As a result, however, of the lack of funds to resettle and police such vast areas, governments have left the *Caiçaras* within protected areas undisturbed, although prohibiting their subsistence activities (primarily swidden agriculture and hunting). Overall, this ambiguous situation has left many families in an unbearable situation, where cases have been reported of households being searched by forest wardens checking cooking pans for evidence of illegal game concoctions.

In this chapter I demonstrate that historically *Caiçara* societies have been underprivileged, both ecologically and socio-economically at global, regional and national levels. Holding a peripheral position within Brazilian society, their identities are often misappropriated for political objectives. My argument here is that the 'ecologically noble' *Caiçara* (mis)representation,

Figure 3 Brazil, showing the Amazonian and Atlantic Rainforests and the Transama-
zonica

although constructed with the aim of assuring their right to remain within
protected areas, has only contributed to their disempowerment. Indeed, the
appropriation of the 'ecologically noble' representation by the *Caiçaras*
themselves, as a political instrument to try to overcome pressing problems,
may trap them forever in a peripheral situation.

Finally, it is suggested that if anthropology is to play a key role in the
debate concerning ecological underprivilege, it should avoid the pitfalls
of synchronicity, examining other societies and cultures in a more processual way (Scoones 1999, Wolf 1997). Anthropology's marginal interest in
Brazilian 'historical peasant societies' (Nugent 1993) is clear from the
scant number of ethnographic case studies of *Caiçaras* (and current representations of the *Caiçaras* perpetuate an image of timeless stasis).
Giving back their histories to peoples not only helps them create their
own identity, thus preventing external manipulation, but also stresses
their role in building natural landscape. This chapter is an initial attempt

to reframe the *Caiçara* case within the context of environmental history (Balée 1995, Hill 1996, Scoones 1999).

A diachronic point of view

The first Brazilians – namely *mamelucos* – were the result of genetic interadmixture of Portuguese colonisers and the several *Tupí-Guaraní* groups which formed a complex and dynamic social system stretching along the coast from southern Brazil to the mouth of the Amazon river, in the beginning of the sixteenth century. The spread of Old World pathogens, massive slave raiding and slavery soon caused the indigenous populations to flee inland, abandoning their villages, which were appropriated and duly became the first European settlements (Wolf 1997). Even if population was drastically reduced, *Tupí-Guaraní* knowledge of subsistence in the Atlantic Rainforest remained key for the survival of the settlers (Schaden 1994). In the first four decades of colonisation (1500–1540) the influence of African slaves was very small, but their subsequent incorporation into the local socio-economy gave birth to what Ribeiro (1995) calls 'the Brazilian people'.

The relationship between pre-colonial Amerindian society and subsequent neo-Brazilian peasant societies (*Caiçaras* and *Caboclos*) has been complex (Nugent 1993). As argued by Nugent, 'instead of there being a focal culture/identity formation process, there is a diffuseness' (1997: 37). For this reason, *Caiçaras* may be considered as *Caipiras* (the peasants established in the south-eastern Brazilian hinterland) by some (Mussolini 1980, Noffs 1988, Pierson and Teixeira 1947, Silva 1993), while others prefer to attribute them a separate identity (Setti 1985).

The identity-formation process of the *Caiçara* merged several cultural features of the Amerindians of the coast (fishing, manioc cultivation and use of native plants) with Portuguese ones (language, Roman Catholic religious practices and folklore). African culture survived only to a limited extent. Several socio-cultural elements link the *Caiçaras* from the littoral to the *Caipiras* from the hinterland: swidden agriculture,[5] voluntary collective work (*mutirão*), labour exchange (*troca dia*), manioc (*Manihot* sp.) flour cultural complex, and Roman Catholic religion (Willems 1996, Mussolini 1980).

The word *Caiçara* has its origin in the *Tupí-Guaraní* term *caá-içara*, meaning 'the man from the coast' (Sampaio 1987). The *Tupí-Guaraní* used this term to refer to the wooden fences that protected households and villages, and also to the sticks used as a device to trap fish on the shore. Later, it began to be used to refer to the huts built by fishermen on the beach, in which to keep their canoes and fishing devices. Eventually it was used to identify the inhabitants of Cananéia, in the south of São Paulo and subsequently all individuals and communities along the coast of São Paulo, Rio de Janeiro, Paraná and Santa Catarina states (Fundação SOS Mata Atlântica

1992). It is not clear exactly when the term *Caiçara* was first used in literature. The first anthropological studies about these societies already referred
to them as *Caiçaras* (Willems 1966, Mussolini 1980) although they make no
reference as to how the Caiçaras referred to themselves.

During the first decades of colonisation, huge amounts of resources
were extracted from the Atlantic Rainforest, mainly brazilwood (*Caesalpinia
echinata*) (Por 1992, Dean 1996). Soon, however, Portugal was faced with the
task of finding an economic substitute for extraction so as to justify the
huge costs of defending its possession. Thus it was that around 1520, Portuguese colonists started producing cane sugar for export in plantations
(*engenhos*) set up in the tropical lowlands of the Brazilian littoral owned by
Portuguese immigrant families. As a result, the colony became definitively
tied to European markets by the production of export crops and by the
expanding commerce of slaves (Dean 1996).[6]

The economic cycle based around sugar production inaugurated in
Brazil a pernicious clientship that would perpetuate itself during the next
centuries and remains embedded in today's political life-style (Valença
1999). The small neighbouring freeholds that supplied *engenhos* with primary goods and other services were linked to the landowners by a
patron–client relationship that reinforced their peripheral position in local,
national and international economies (Furtado 1968, Wolf 1997). While
the patron (*patrão*) provided resources, protection and links to the outside
world, the client had to offer support and obedience: 'The *engenhos* were
under the landowners' control and few dared to challenge them, not even
Governor-Generals and Bishops. The boss's wealth – land, slaves, mills and
other equipment – and their strategic role in Portuguese colonial policy
made them increasingly powerful' (Valença 1999: 6).

The small *Caiçara* communities that settled around the *engenhos* in the
sixteenth century were also greatly influenced by the landscape, particularly
along the south-eastern coast. The high altitudes of the south-eastern
coastal mountain range (*Serra do Mar*) covered with Atlantic Rainforest and
the little availability of land for agriculture in the coastal plains restricted
the number of *engenhos*, and also forced the *Caiçaras* to settle in small coastal
plains and piedmonts, where they produced manioc flour and fish (Dean
1996, Mussolini 1980). Thus, besides occupying a peripheral political economic position, *Caiçaras* were also pushed to marginal lands in the slopes of
the Atlantic Rainforest. Most of all, by limiting the size and mobility of
Caiçara communities, geographical features influenced their manner of
landscape occupation and use of natural resources, as well as their social
morphology (França 1954, Mussolini 1980, Marcílio 1986).

Colonial Brazil was characterised by a sequence of economic cycles.
After the Dutch invasions in north-eastern Brazil upset the sugar economy
in the first half of the seventeenth century, Portuguese investments focussed
alternatively on tobacco, mining, cotton and coffee (Furtado 1968).

In the south-eastern littoral of Brazil, periods of intense economic activity alternated with periods of stagnation, population decrease, simplification of subsistence activities and increase in the number of small communities.[7] In 1836, the northern coast of São Paulo State had 370 coffee farms, twenty distilleries of *cachaça* (liquor made from cane sugar, similar to rum) and seventeen sugar mills. In the southern coast of São Paulo State, on the other hand, rice cultivation prevailed, due to existing ecological conditions. Extensive, well-irrigated coastal plains allowed for the existence of eighty-two rice processing plants (Willems 1966).

As shown by França (1954), in São Sebastião Island (northern coast of São Paulo State) demographic density fell from 32.7 inhabitants/km² in the middle of the nineteenth century to 14.3 in 1950, although 86 percent of the population still lived exclusively from subsistence activities. In the same period, the amount of land undergoing cultivation fell from 126 to 6.8 km², contributing to the regeneration of the Atlantic Rainforest. Indeed, 'changes on the level of the world market had consequences at the level of household, kin group, community, region and class'(Wolf 1997: 310).

Until the 1930s to 1950s the *Caiçaras* could be considered mainly as shifting cultivators and coastal canoe fishermen.[8] This started to change when a few Japanese fishermen who relied on motorboats and fishtraps (*cerco*) for outstanding gains in production settled in the region. Several authors describe the effects of these two innovations. As male labourers shifted to fishing, agriculture was partially or totally abandoned. Thus, *Caiçara* way of life was reorganised and the sea acquired a significant new status in their culture (Mourão 1971, Silva 1993).[9] According to Willems, *Caiçaras* were characterised by 'a general alertness to new economic opportunities which are readily embraced or readily rejected depending upon price levels; considerable spatial mobility which is manifested in the frequent migrations of individuals to Santos and in trading expeditions to rather distant localities in the coast' (1966: 6).

Thus, rather recent features of their lives were appropriated as central facts through which the *Caiçara* could be imagined as 'ecologically noble', that is, as traditional fishermen, familiar with a maritime symbolic and technological world. This process of hiding the agricultural past of indigenous peoples as a strategy to try to enhance their 'nobleness' is hardly exceptional, as Novellino (this volume) demonstrates for the Batak of the Philippines.

During the same period, wealth generated by the coffee economy in the south-east launched Brazilian industrialisation, giving rise to urban middle classes. After the 1950s several new highways were built to link littoral to hinterland, and socio-economic transformation accelerated. Urban middle-class tourists came to buy land from the *Caiçaras*, who moved inland to the fringes of the Atlantic Rainforest or to nearby growing towns. Although they had informal rights over their land (*usocapião*) most *Caiçaras* did not have registered legal property. It was not unusual for real-estate entrepreneurs wishing to occupy large amounts of land to threaten *Caiçaras* who resisted moving out (Silva 1979).[10]

Tourism and urbanisation quickly transformed the rural landscape, and *Caiçara* agricultural subsistence definitely lost its economic importance. As they abandoned their dwellings they also lost the easy access to the beach, and so coastal subsistence fishing was also gradually abandoned. Subsequently, in the 1980s, when the government began to establish restrictive protected areas in Atlantic Rainforest remnants, prohibiting swiddening cultivation and hunting, *Caiçara* families that still relied partially on agriculture had to change to artisanal fishing and services for tourists in order to survive (Luchiari n.d., Vitae Civilis 1995).

The 'ecologically noble' Caiçara

Brazilian protected areas have historically been established for scientific and aesthetic reasons, the main goal being to isolate great areas from human activity, with exceptions made for research and tourism (Vianna 1996). One of the most illustrative examples of the deleterious effects of this kind of policy on local people was the establishment of the Juréia Environmental Station (*Estação Ecológica da Juréia*) to protect part of the Atlantic Rainforest's remnants. It is the starting point of the debate over *Caiçaras'* right to remain in their ancestral lands.

In 1979 the Federal Environmental Secretariat (SEMA) signed an agreement over an area of 1,100 ha in the south of the State of São Paulo, establishing the Juréia Ecological Station. SEMA's director was a biologist and conservationist, very much concerned with Atlantic Rainforest biodiversity. In his own words, one of the main 'problems' to be solved was the existence of *Caiçara* settlements within the limits of the ecological station. Although admitting that these people deserved respect, he considered that 'future generations have priority in relation to present necessities'. Care should be taken for local people's activities not to interfere with the main goal of the ecological station, that is 'to allow scientific research and protect natural ecosystems and biodiversity' (Nogueira-Neto 1991, my translation). In 1986 the State of São Paulo government increased the protected area to its present limits (72,000 ha) encompassing twelve *Caiçara* settlements within the Juréia-Itatins Ecological Station. Environmental civil servants working in close contact with the *Caiçaras* were divided into two groups, one concerned over the legal impediments to their livelihoods, the other anxious that their traditional activities damaged the environment, due to *Caiçaras* 'low scholarly/educational, cultural and technological level' (da Costa 1991, my translation).[11]

Influenced by the overall political setting of the first years of the 1980s (the end of military government) and by international environmental conservation movements, the first environmentalist non-governmental organisations (NGOs) aimed at protecting the Atlantic Rainforest were founded.[12] These movements called attention to the many external pressures that incurred upon the Atlantic Rainforest's remnants (hunters and extractivists, real-estate agencies and tourists) and launched the debate over its conservation and over the existence of *Caiçaras* living within protected areas.

As theoretical/authoritative evidence was needed to corroborate the position of those championing the rights of *Caiçaras* to remain in their own land, sociologists and anthropologists became involved in the debate. However, with less scholarly and/or empirical input, and with rather more political objectives, what started to be emphasised was the close relationship of the *Caiçaras* with nature. The 'ecologically noble' representation of the *Caiçara* is portrayed in scientific literature by several authors (Diegues 1988, Cunha and Roguelle 1989, Milanelo 1992, Diegues 1993, Diegues and Nogara 1994). This was to be a counterinfluence to the discredited human vs. nature dichotomy adopted by conservationists (see also Halder this volume).

However, though criticising the human vs. nature dichotomy, these academics implicitly used arguments which relied upon the very same dichotomy to create a 'naturalised' *Caiçara*. The 'traditional' (natural, isolated, pristine) communities, claiming the right to stay inside protected areas, were opposed to 'non-traditional' ones (human, acculturated, modern) thus reinforcing the human vs. nature dichotomy instead of overcoming it (Descola and Pálsson 1996a, Ellen 1986, Scoones 1999). By emphasising the 'naturalness' of *Caiçara* societies, these authors deprived them of their past, creating an ahistorical society possessing a static culture.

In the literature, as well as in common usage, *Caiçaras* are referred to as 'traditional people' (*populações tradicionais*), a very broad and generic designation of Brazilian rural societies. However, the concept of 'traditional people' in Brazil lacks scientific exactness and is based more on the attributes of Amerindian groups than on those of peasant societies (Vianna 1996). The *Caiçaras* are described as small isolated groups living in the coast of São Paulo, Rio de Janeiro, Paraná and Santa Catarina states (Willems 1966) who rely on natural resources to survive and live in 'harmony' with nature. This 'harmonious' relation is usually based on the fact that they live in the few preserved Atlantic Rainforest remnants, which is taken as proof of their conscious secular management of the environment. In other words, *Caiçaras* have become idealised as 'primitive, harmonic, symbiotic and conservationist' by part of the environmentalist movement, by environmental civil servants working with protected areas, and by the media (Vianna 1996: 210, my translation).

Caiçaras are not unique in this sense. Rather, they form part of the mythology of late-twentieth-century environmentalism that contrives to perpetuate a 'pernicious dichotomy, which, however much some have tried to disguise it, reproduces the notion of a primitive, exotic Other' (Ellen 1993: 126). The Batak of the Philippines (Novellino this volume), and many others have also undergone the same process of cultural redefinition.

My point is that what gave birth to the *Caiçaras*' contemporary situation, their reduction to 'stereotypic symbols of isolation and alienation' (Hill 1996: 9) from national society, has been their removal from colonial, national and global histories. This only contributes to their disempowerment. Regrettably, anthropology has contributed very little to deconstructing the negative identity of *Caiçaras* as backward *mestizo* peasants, an image repackaged in modern terms as ecologically virtuous.

I propose that the concept of ethnogenesis (Hill 1996) offers an underlying analytical tool for developing a critical historical approach to *Caiçara* culture. In Hill's definition, ethnogenesis is a 'synthesis of a people's cultural and political struggles to exist, as well as their historical consciousness of these struggles' (1996: 2). Although the importance of a historical approach to indigenous cultures as well as the unrealistic possibility of isolation and autonomy of Amerindian societies has already been addressed in the Amazon (Hill 1996, Leonard 1999, Riviére 1993, Whitehead 1993a,b) the *Caiçaras* are still addressed as static local cultures.

To use the concept of ethnogenesis some adaptations have to be made, so as to adjust it to analyses of *Caiçara* societies. For instance, I am looking at the ethnogenesis of a new cultural identity that appropriated features from several *Tupí-Guaraní*-speaking groups, but also Europeans (mainly Portuguese) and, to a lesser extent, from different African societies. Secondly, although *Caiçara* societies are a *result* of violent changes imposed upon the *Tupí-Guaraní* and other Amerindian groups, 'including demographic collapse, forced relocations, enslavement, ethnic soldiering, ethnocide and genocide', their existence is also marked by a process of 'creative adaptation to a general history of [...] changes' (Hill 1996: i)

The use of the ethnogenesis concept enables one to include the local history of *Caiçara* societies within the complex interrelations of global history (Hill 1996, Wolf 1997). By breaking the constructed static image of *Caiçara* culture I hope to bring back their dynamic history and cultural adaptive richness, as a means of empowering their political struggles for their rights to remain on their traditional lands.

The historically underprivileged *Caiçara*

Caiçaras are a result of the expansion of European power and trading activities that started to change the world irreversibly by the end of the fifteenth century, and have historically been underprivileged, both ecologically and socio-economically. In this sense, the *Caiçaras* as well as the Amazonian *Caboclos* and other Latin American *mestizos* are the outcome of long-term processes of colonial expansion and globalisation, now dressed as eco-colonialism (see Nygren and Nugent this volume).

Being descendants of people with no civil or political status (Indians, African slaves) the free *mestizos* in general, and the *Caiçaras* in particular, had no place in the strongly hierarchical Brazilian colonial society, divided into two social classes: landowners and slaves (Chaui 2000).

Ever since the sixteenth century, *Caiçaras* have played a peripheral economic role in much wider political and economic systems (mercantilist and capitalist) (Willems 1966, Wolf 1997). They were subjected to different owners of political power such as local patrons (Furtado 1968, Valença 1999), entrepreneurs (Campas 1980, Silva 1979) and urban middle classes, occupying peripheral lands which they did not legally own (Dean 1996, Mussolini 1980).

Up to the beginning of the twentieth century, *Caiçara* communities were left to stagnate within the Atlantic Rainforest, together with the also underprivileged *Quilombos* and remaining *Guaraní* tribes, as cyclic bursts of economic development happened elsewhere. When the south-eastern coast of Brazil was rediscovered by urban middle classes in the 1960s, they began to lose their land, as they were unable to cope with economic and sometimes coercive power. Finally, when protected areas began to be established in the Atlantic Rainforest in the 1980s, they had to struggle against the government and conservationists. Ironically, the holders of political power were seeking to protect the same landscape *Caiçaras* helped to create (Oliveira 1999).

Since there were no formal political organisations within *Caiçara* societies, many different groups were formed, each one claiming to be their representative.[13] As a scientific base to their claims was desirable, members of the academic community became involved. In turn the *Caiçara* struggle was a perfect way for the owners of scientific knowledge to legitimate their own concerns (Adams 2000a, Milton 1993, Vianna 1996). As Ellen (1986) points out, far from being in 'harmony' with nature, 'traditional' societies are often cruelly the victims of it.

I argue that the misappropriation of *Caiçara* identity by partisan sociologists and anthropologists only helped to detach them from their social and historical context, perpetuating their position as poor survivors of a distant past, only able to make a living out of simple subsistence practices, within the remnants of a formerly luxurious Atlantic Rainforest. Present Brazilian society is still marked by its colonial past and still has a highly hierarchical structure. By means of a process of naturalisation of social inequalities (social inferiority is considered a natural status for women, Indians, etc.), the historical genesis of difference and inequality is suppressed, allowing for the existence of visible and invisible forms of violence, which are not perceived as such (Chaui 2000).

In fact, even the use of the word *Caiçara* by sociologists and anthropologists was, in itself, a way of creating a boundary between these traditional societies and modern national society, as historically it has had a pejorative meaning. As observed by Willems (1996), in the first half of the twentieth century, the word *Caiçara* always invoked the underdeveloped, backward, impoverished and malaria-stricken southern coast of Brazil. The stereotyped *Caiçara* was considered a lazy, immoral drunkard, indolent and unreliable. Besides, Setti (1985) observed that although the inhabitants of Ubatuba, on the north coast of São Paulo State, admitted being called a *Caiçara* they did not define themselves as such. On the contrary, they called themselves *ubatubanos* (born in Ubatuba), *praianos* (from the beach) or *barriga verde* (green belly), because of the abundance of green bananas in their diet, eaten with fish and manioc flour.

However, a historical analysis shows that distinctions between traditional and non-traditional, rural and urban, pristine and acculturated, isolated and modern do not fit the *Caiçaras*. The diversity of subsistence strategies adopted by the *Caiçaras*, not always directly related to nature, demonstrates their ability to adapt to an ever-changing economy, as can be exemplified by the arrival of the motorised fishing boat in the beginning of the twentieth century (Adams 2000b). The 'isolation' to which *Caiçara* communities were subjected before the arrival of the motorboat and the construction of highways (1950s–1960s) was always relative and associated chiefly with periods of economic stagnation (Willems 1966, Mourão 1971, Setti 1985, Silva 1993). Indeed, during periods of prosperity when other economic alternatives were offered to the *Caiçaras*, subsistence agriculture and fishing were abandoned. On the other hand, in times of economic stagnation they returned to the rural areas and to their adapted subsistence activities (Mourão 1971).[14]

As observed by and Mussolini (1980), *Caiçara* subsistence activities could even result in an 'intense, exclusive and even abusive usage of natural resources'; 226, my translation), which does not agree with their modern ecological (mis)representation.[15] As an adaptive society, integrated into both urban and rural environments, the *Caiçaras* raise important issues for the anthropological characterisation of subsistence activities.

Conclusion

The 'ecologically noble' *Caiçara* is a misrepresentation constructed for political ends and should be abandoned and replaced by a dynamic and historical definition of *Caiçara* identity. As shown elsewhere, the dichotomy between 'traditional' and 'non-traditional' people 'cannot be transformed at a later stage into a fully historical understanding of cultural identities' (Hill 1996: 9). The new identity should try to encompass both written history and the *Caiçaras* own collective memory. Although the first publications about *Caiçaras* were made in the 1930s, no ethnographic work was carried out until today and this is a gap that has to be filled by anthropology.

As has already been shown for São Sebastião Island (Adams 2000a) and Ilha Grande (Oliveira 1999), the *Caiçara* debate would greatly benefit from historical analyses of 'people in places', considering the 'environment as both the product of and the setting for human interactions' (Scoones 1999: 490). In fact, in some areas the remaining 'pristine' Atlantic Rainforest is, in reality, highly influenced by historical *Caiçara* management (Oliveira 1999).[16] The synchronic image of traditional fishermen only withholds the importance of the *Caiçaras* in the construction of the landscape that is now under governmental protection. Furthermore, the historical approach might be the missing link joining biology and anthropology to investigate human societies and their natural environments.

If anthropology is going to play a key role in the debate about ecological underprivilege, it should avoid the pitfalls of synchronicity. In order to do so, contributions from other disciplines such as environmental history and ethnobotany, as well as the concept of ethnogenesis, should be considered. Only by examining other societies and cultures in a more processual way shall anthropology be able to overcome the human vs. nature dichotomy.

Acknowledgements

I would like to thank Dr Eeva Berglund and Dr Stephen Nugent from the Anthropology Department of Goldsmiths College, University of London, for their kind invitation to present this paper. I am also grateful to Dr Roy Ellen and Dario Novellino (University of Kent at Canterbury) for their revisions and comments.

Notes

1   Federal Laws 4.771/65, 5.197/65, 6938/81 and 6902/81.
2   Nogueira-Neto, P. 1991. Personal communication.
3   But also some *Guarani* tribes and *Quilombos* (communities established by runaway African slaves).
4   For a detailed description of the Atlantic Rainforest and its deforestation process see Mantovani (1990, 1993) and Dean (1996). For a summary see Por (1992).
5   A review of the agricultural system of the *Caiçaras* in the twentieth century (Adams 2000c) showed that the size of the gardens has been diminishing since the 1950s. Average size of gardens is 0.42 ha, average period of cultivation 3.1 years, and average fallow period 7.8 years.

6   The raiding *Bandeirantes* of São Paulo alone supplied sugar plantations of north-eastern Brazil with 350,000 slave Amerindians during the slavery period in Brazil, which lasted until 1750. The Amerindians were later replaced with African slaves brought mostly from Congo, Angola and Mozambique. Between 1811 and 1870 Brazil acquired as many as 1,145,000 African slaves (Wolf 1997).

7   This scenario has been described for Paraná (Cunha and Rougeulle 1989, IPARDES 1989, Langowiski, Simão, and Goldman 1958, Rougeulle 1989) São Paulo (Almeida 1946, França 1954, Goldman 1958, Willems 1966, Silva 1975, Mussolini 1980, Diegues 1983, Winther et al. 1989, Silva 1993) and Rio de Janeiro (Siqueira 1989).

8   Some communities depended mainly on fishing (Bernardes and Bernardes 1950).

9   Conversely, not all communities adapted well to the changes brought about by the motor boat, as shown by Rougeulle (1989) on the coast of Paraná. In fact, some communities never made the transition, such as Mandira in the south of São Paulo State (Sales and Moreira 1994).

10 In Trindade, a small community on the coast of Rio de Janeiro State, local peo-ple were threatened in the 1970s by the multinational holding Brascan-Adela, which had acquired a huge amount of land nearby for a real-estate enterprise. Wishing to add more land to their plot, Brascan-Adela used lawyers, local author-ities, and dozens of armed men to intimidate people and force them to leave their houses (Campos 1980).

11 State of São Paulo Environmental Secretariat, internal report.

12 Only in the 1980s did Brazilian environmentalism become a multisectoral move-ment, involving environmental government agencies, environmentalist community groups, social movements, scientific environmental researchers (including social scientists), and a few private enterprises (Viola and Leis 1995).

13 Only in 1994 was an association of inhabitants of protected areas founded (*União dos Moradores das Unidades de Conservação*) (Vianna 1996).

14 Neglect of important features associated to their past as cultivators distances the *Caiçaras* from their *Caipira* common ancestry as defined by Cândido (1964), reducing their cultural richness (Adams 2000b).

15 Besides, some descriptions of *Caiçara* mismanagement of the Atlantic Rainforest and maritime resources have already been reported, such as erosion, exposed rocks and soil, unnecessary burning (França 1954, Willems 1966, Mussolini 1980), and overfishing (Sales 1988), suggesting that the 'ecologically noble' image does not fit reality.

16 As has already been shown in the Amazon for other societies (Adams 1994, Balée 1987, 1992).

2

NATURE AS CONTESTED TERRAIN
Conflicts Over Wilderness Protection and Local Livelihoods in Río San Juan, Nicaragua

Anja Nygren

Introduction

> I remember when MARENA [Ministry of Natural Resources] came here to control the
> forest fellings, I cut down my whole forest for pasture – bam, bam, bam. Seventy *man-*
> *zanas*, whruum! I cleared all the land even close by the river. So when MARENA came,
> they didn't catch me. You know, if they want to control us, we fell our forests. But if
> they give us a chance to live on something, we aren't going to cut down the trees.

This piece of history was narrated by don Anastasio,[1] a small-scale forest extractor from Boca de Sábalos at a workshop on non-timber forest products
organised in Río San Juan, Nicaragua, in 1997. Short and pithy, it reveals one
of the most problematic issues in the current debate revolving around competing claims over tropical forests. In this heated arena of discussion, many
conservation authorities portray the tropical forests as irreplaceable sanctuaries
of biodiversity and natural scenery that should be preserved free from human
interference. This view, which satisfies national and global environmental agendas, is criticised by many Third World smallholders, according to whom the
protection of tropical forests as untouched areas of wilderness has no justification if it violates the local rights of access to productive resources (Ghimire and
Pimbert 1997, Neumann 1998).

  This chapter analyses the conflicts between local settlers, non-governmental organisations (NGOs) and state authorities over access to natural
resources in a protected area buffer zone of Río San Juan, in south-eastern
Nicaragua. It shows how control over natural resources is defined and contested within the political arenas at different levels, from localities and regions

to national and international institutions (cf. Peet and Watts 1996, Ribot 1998). The study analyses the struggle over the forests in the buffer zone of the Indio-Maíz reserve by examining the multiple discourses on nature protection and local livelihoods. The main aim is to reveal how the local settlers have been characterised in the wider politics of environment and development, and how these representations have been experienced by the settlers themselves. Such a perspective offers interesting angles from which to study the complexity of power structures mediating the social relations of natural resource utilisation and the ebb and flow of competing environmental perceptions (for a parallel problematique in Brazilian Amazonia, see Adams in this volume).

The chapter begins with a brief description of the environmental and social landscape of Río San Juan. Then, the local environmental conflicts are examined as a multifaceted struggle over resources and representations, involving contested claims of who has the right to make decisions concerning the local resources. By connecting the environmental conflicts of Río San Juan within the corresponding processes and discourses at regional, national and global levels, the study aims to shed light on wide-ranging debates over nature protection and social justice, and control and authority. As will be shown in the following analysis, diverse social actors, such as local smallholders, forest extractors, absentee cattle-raisers, conservation authorities, development experts, NGOs, transnational companies and international aid agencies make competing claims over natural resources in Río San Juan. All these actors suggest their own solution to the local nature-based conflicts and in this way attempt to strengthen their power in the control over resources and environmental images.

The forest-edge communities of Río San Juan: Diversity of actors and policies

Inspired by the ideology of nature protection and by international concern over tropical deforestation and loss of biodiversity, Nicaragua is transforming much of its remaining forests into protected areas. The idea of tropical forests as a 'threatened home of endangered species' and a 'natural patrimony for future generations' has provided a powerful justification for wilderness protection in Nicaragua, as elsewhere in the tropics, although there are also those who question the policy of allocating as much land area as possible to regimes of protected area management.[2] In 1997, 18 percent of Nicaragua's total land area had been set aside as protected areas (Segura et al. 1997).

The biological reserve of Indio-Maíz, located in the humid tropics of the department of Río San Juan, is one of the biggest protected areas in Nicaragua, covering 264,000 hectares of land.[3] The reserve, which was established in 1990, has acquired an international reputation as one of the most outstanding protected areas in Central America. It belongs within the category of strictly protected areas; the only activities permitted inside the reserve are scientific research and wilderness protection (IRENA 1992).

The establishment of the Indio-Maíz reserve has many implications for the livelihood opportunities of the surrounding forest-edge communities. The buffer zone of the reserve in the municipality of El Castillo covers 180,000 hectares of land and has some 15,000 inhabitants, who live in nine settlements and dozens of smaller villages (DANIDA 1998). The region belongs to one of the most intensive agricultural frontiers in the country, with high rates of immigration and deforestation. To secure local support for the protection of Indio-Maíz, many rural development projects have been initiated in the buffer zone. In 1994-1998, there were thirty projects underway in Río San Juan with a total budget of US$21 million, involving agricultural diversification, community forestry, ecotourism, environmental education, non-timber forest products and women in development, with financing from the United States Agency for International Development (USAID), the International Development Bank (IDB), the European Union, the Danish International Development Agency (DANIDA), Germany's Deutsche Gesellschaft für Technische Zusammenarbeit (GTZ), and various international NGOs (Vegacruz 1995). Most of the projects were implemented by Nicaraguan state institutions, NGOs and local municipalities.

Until the 1950s there were scattered hamlets of smallholders in the buffer zone. These households cleared small patches of forest for crop production, and they also practised hunting and gathering. The extraction of rubber, chicle (*Manilkara sapota*), wild animals and precious timber species formed an important part of the local livelihood strategies. During the 1960s and 1970s, a wave of new colonists entered the region, principally smallholders from Pacific areas who had lost their lands to cattle estates and cotton plantations. These 'people without land' searching for 'land without people' began to open up the forests of Río San Juan to slash-and-burn agriculture. According to the agrarian legislation of the time, those who 'improved' the land through forest clearing acquired a perpetual right to the area they had cleared (Maldidier and Antillón 1996). Since the decline of the extraction of rubber and chicle in the 1940s and 1950s, the extraction of raicilla (*Psychotria ipecacuanha*) became an attractive livelihood strategy until its price began to decline dramatically in the late 1970s.[4]

The Nicaraguan civil war of 1979-1990 largely depopulated the region. Most of the people left as refugees for Costa Rica or they were evacuated to government-established settlements (*asentamientos*) located in the more controllable areas of the municipality. In these settlements, agricultural production was organised through cooperatives, which had access to state-owned land, credit and assistance (Utting 1993: 147–150). The socialist Sandinista government put major emphasis in rural development policies on land redistribution and increased agricultural production, while the environmental issues were often left with relatively scarce attention (in this respect, cf. the essays by Anderson and Rolle in this volume).

Since 1990, a considerable number of the refugees and internally displaced people have returned to their farms in the interior. At the same time, the flow of new colonists entering the region has dramatically increased. Most of them come from the cattle-raising area of Chontales where there is no free land left for cultivation. As a compensation for making peace, the Chamorro government donated areas of land in the buffer zone to those who had occupied a high rank in the Sandinista or Resistance army during the war. Many of these demobilised groups were given ownership of the land already in the possession of the smallholders. As they had no title to the land, the smallholders were unable to file legal claims to their possessions. The result is a high level of land and natural resources conflicts with varying degrees of violence.

Most of the current inhabitants are non-indigenous peasant farmers (*campesinos*); the region's indigenous population was largely exterminated or dispersed during the eighteenth century (Rabella 1995). These *mestizo* smallholders cultivate maize, beans and rice by slash-and-burn agriculture and supplement their livelihood with small-scale forest extraction, logging and trading. Many of them also participate in two-step migration, which involves clearing land for pasture and then selling it to land speculators. There is a high degree of mobility; people come and go, and many of them move ever further into the hinterland. The agricultural frontier is now reaching the boundaries of the Indio-Maíz reserve.

A majority of these smallholders are struggling to survive in a situation where the access to free land has ceased, crop productivity is low, and high transportation costs and hierarchical forms of commercialisation make it difficult for small-scale producers to compete in markets. The ongoing structural adjustment policies have only increased the economic hardships of many smallholders. As a consequence, people move between spheres where different economic opportunities seem available and the informal sector plays a crucial role in this frontier economy.

All this has provoked a series of conflicts between the forest-edge settlements, conservation authorities and development advocates working in the region. Many settlers bitterly explained that during the Sandinista government, the land was distributed to 'those who really depend on earth for living, while today, nature is preserved for its own sake'. In a survey completed by the Ministry of Natural Resources (MARENA) in 1998, three hundred families were recorded as squatting inside the Indio-Maíz reserve. Most of them were classified as 'invaders' who had already benefited from land redistribution in another region but who had then sold their land and moved further into the reserve. One third of these families left the reserve peacefully due to governmental pressures, and the rest were to be evicted by the Nicaraguan army (*La Tribuna* 16.04.1998). However, after a series of violent confrontations provoked by the situation, the government decided to reduce the area of the reserve, by excluding the 31,000 hectares of land under invasion (*La Gaceta* 18.06.1999).

All this diversity of actors and interests involved in the natural resource conflicts on this forest frontier makes the whole struggle over sustainability and environmental equity extremely complicated. The main information presented here is based on my ethnographic field research in Río San Juan in 1996-1998. The primary data consists of ninety hours of tape-recorded interviews and participant observation involving the local people and forty-five conservation organisations, development institutions and NGOs. This material was supplemented by existing archive and statistical material, ministerial documents, law texts, development reports and historical documents (see Nygren 2000).

Contested claims of authority

Experts planning for the local people

One of the principal questions in the debate over natural resources in Río San Juan revolves around the issue of who holds the power to control access to them. According to many conservation authorities, it is the task of the state to control the 'national heritage' of Indio-Maíz through improved vigilance. These officials also stressed the need to enforce permit control in the buffer zone; they argued for increasing restrictions on local people's access to forest products and for more efficient sanctions for unauthorised forest clearing. Many officials were of the opinion that local people exploit the forests in order to make profits, rather than because of a lack of alternatives to meet their basic needs of living.

Given this conservation agenda, reserve managers showed little sympathy for local claims to resource extraction. Most of them held the view that the local extractors encroach on the forests with little environmental awareness. They also questioned the development advocates' suggestion that local people participate as guards of the reserve. According to them, it was only necessary to give the reserve managers the authority to make the people respect the law. There were also authorities who argued that the only possibility to save Indio-Maíz would be the utilisation of the 'hard hand' (*la mano dura*). They supported the forced eviction of the squatters inside the reserve, and they also had plans to use a 'green army', composed of Nicaraguan armed forces, to patrol the reserve. All this was justified by the arguments of the local 'culture of violence', in which the only recognised law was seen to be the 'law of the jungle' (*la ley del monte*). According to these officials, 'as everybody on this forest frontier can use a rifle and is accustomed to appropriate whatever he wants without the presence of the army, people will invade the whole reserve'.[5]

The authorities of the National Institute of Agrarian Reform (INRA) defined the task of the state in that region to be its stabilisation. According to them, the primary need in this zone of 'spontaneous' colonisation was to rationalise the chaotic land-use patterns by 'ordering the disordered', and registering all the settlers and farmlands in the institution's archives. Such property registration was then proposed to stimulate more consciousness of conservation. There was little indication, however, that given the existing complexity of land and resource tenure, the institution's ambiguous land registration programme would state explicitly which rights are secured for whom.

Not all the state authorities supported such a rigid agenda of controls. The authorities at the Ministry of Economy and Development (MEDE) criticised the view of forest and land registration officials as too paternalistic. In their view, the concept of untouchable protected areas is outdated; instead, the door has to be left open for ecotourism and bioprospecting, and for the increasing participation of the private sector in natural resource management, because the sustainable marketing of natural resources is one of the few possibilities Nicaragua has to improve its future prospects. The authority to control this biobusiness was to be given to MEDE; among these authorities of the central government there was little enthusiasm for decentralisation.

The strengthening of local governments is nevertheless an increasing tendency in Nicaragua, a trend which is being strongly supported by international donors. Through decentralisation, local people are supposed to develop a better sense of ownership of the rules regarding resource use and be more inclined to obey them. There is, however, little empirical evidence as to whether decentralisation is good for the forests and the people who depend on them. In the worst scenario of Río San Juan, this could lead to local governments becoming even more vulnerable to political pressures from regional powerholders. There is currently intensive lobbying by timber companies, African oil palm entrepreneurs, cattle-raisers and a transnational mining company to pressure the municipal authorities into releasing the buffer zone of Indio-Maíz from the strict conservation agenda. Moreover, a vital role in the politics of Río San Juan has always been played by a handful of local *caciques*. These mainly older, male settlers act as invisible political figures behind the collectively elected local committees of development, controlling the flow of benefits and access to resources in their communities. Without careful consideration of local power structures, decentralisation may lead to *caciques* taking the power to redefine development projects and determine those who are to benefit from them.

Government decentralisation is currently paralleled by NGO-based development in Nicaragua. Most of those NGOs working in Río San Juan challenged the state's coercive conservation policies by pointing out that nature protection has no future if the livelihood requirements of the local inhabitants are not taken into account. According to many of them, the state has proved itself unable to manage protected areas and to improve the living conditions of the rural poor in Río San Juan, as elsewhere in Nicaragua. In their view, it was time for the government to agree to the key role of NGOs as promoters of development, because of their better methods to empower local people as active partners in conservation.[6]

In reality, many of these NGOs demonstrated the same lack of attention to local development needs as the government projects. By defining themselves as facilitators of change, most of them seemed convinced that their task is to plan for rural people. The majority of their projects were dedicated to conventional issues, such as agricultural diversification and environmental education, with little attention being paid to the unequal distribution of resources. In order to have a rapid impact, they only worked with landowners, and thus failed to take landless people, squatters and other vulnerable groups sufficiently into account in their efforts. The majority of these NGOs were urban organisations, consisting of educated middle-class members, many of whom worked as state officials in earlier governments. They thus ran the risk of becoming professional, profit-making organisations and thereby losing their legitimacy as civil society actors in the eyes of local populations.

Local reaction and resistance

The main preoccupation among the local inhabitants in the natural resource management of Río San Juan concerned the rights of access to productive resources. In my daily conversations with local settlers, they seriously doubted whether the benefits to be derived from the protection of Indio-Maíz would ever be directed to the local communities. The coercive policies of regulating resource use were seen as a serious hindrance to their ways of living, prohibiting them from practising slash-and-burn agriculture, ordering them to live in registered settlements, and forcing them to apply for permits to undertake any resource extraction whatever.

The same problem concerned the local inhabitants' relations with the development advocates. According to local settlers, the same persons always assume that they have the right to represent the community in negotiations with the development experts, while these experts in turn claim to represent the development needs of the local people in the negotiations with the donors. A common image of the local people among the development advocates was that of 'small-scale farmers', which masked all the social distinctions between men and women, landowners and squatters, poor and more well-to-do settlers. Although the developers claimed to be fully cognisant of the significance of gender in development, their discourse often remained gender-blind, with little awareness that the lines of resource access between men and women depend upon the type of activity, resource and location. There was also little recognition of the fact that local people are simultaneously caught up in different social orders: one being local hierarchies of age, gender, religion, political identity and client-patronage; the other that of the national society and world economy, with their centres and peripheries of power.[8]

The local inhabitants themselves resisted being characterised as 'peripheral slash-and-burn agriculturists'. Although most of them were small-scale cultivators, they also worked as timbermen, shopkeepers, seamstresses, casual wageworkers, midwives, healers and itinerant dealers. There were high rates of intra- and interregional migration, many people working as seasonal labourers in Costa Rica or in urban settings. In this panoply of livelihoods, the developers' portrayals of the homogenous settlers and the generic settler way of life directed attention away from the complex hierarchies of politics and power.

This homogenisation was more than just a reflection of the developers' ignorance of the local social and cultural reality; it also created negative images. These colonist smallholders were generally characterised as 'rootless penetrators who mindlessly destroy nature's precious gifts'. They were repeatedly called *haraganes* (idlers) who are reluctant to work and only take advantage of others if not kept under surveillance. Such was the view of one of the project leaders working in Río San Juan:

> Educating these people for conservation requires much patience because few of them show more than a shallow concern for deforestation. In these forests you're more likely to hear the knocking of an axe than the squawk of a parakeet. These colonists seize a tract of forest to clear it for three harvests of maize, then they sell the plot, and go with their machetes to demolish another patch of forest. It's very difficult that these people, with a vicious cycle of destruction, would show any interest in protecting the forest.

Such portrayals of 'colonists with a pioneer mentality' have affected the local people's self-image and feelings of self-worth. In my conversations with local inhabitants, they repeatedly portrayed their community as a place in 'great need of development' and themselves as 'people of the jungle, who have to wade in the mire to eke out their survival'. Peasantry was a key term of self-description by many smallholders, however it was often an identity associated with shame, referring to something like a 'man of unpretentious beginnings' (*de origen humilde*) or the 'poor people of the interior' (*gente pobre de adentro*). This can be seen in the comment by doña Tina, one of the older settlers in the community of Kilómetro Veinte who remarked: 'Sometimes I feel ashamed when people come here from the outside, and see how poorly we *campesinos* live in these thatched huts and in this lonely jungle.' As has been noted elsewhere (Nugent 1997: 38), these are identities which are constructed in peasants' everyday hardships and in relation to the representations imposed upon them by outsiders.

Local settlers were also well aware of the anthropologists' and other outsiders' romanticisation of 'indigenous peoples'. When explaining my interest in studying the local environmental relations to some inhabitants, they ironically told me that 'here there are no Indians, and no traditional culture, but only *campesinos* who wear polyester pants, faded T-shirts and shabby baseball caps'. Given this experience, it was difficult for these people to imagine that somebody would be interested in their culture and their relationship to nature. Many of them were convinced that 'development' would come to the region from elsewhere and that such improvements as roads, schools and health-care centres would eventually lead to well-being. As a consequence, they often talked of development as things: donations of barbed wire and corrugated-iron sheets or new breeds of chicken and types of pineapple.

This does not mean that people were passively waiting for everything to be done for them. The subjugation of their knowledges rarely went uncontested, and people interpreted the developers' messages in complex ways that helped to bolster their own agency. In our daily conversations, people often insisted that they had no idea of the development projects operating in their communities – even those persons who regularly attended the projects' village meetings. By this rejection, they implied that 'the developers were not trying to resolve their problems'. Many smallholders recalled that they had first been inspired by the projects' promises of local empowerment, but after realising the developers' steadfast intention to regulate access to resources, they ceased to cooperate further. The following comment by don Guillermo, one of the poorest settlers in the community of Las Maravillas, demonstrates that the passivity of the local people was not simply a sign of powerlessness; it was also a strategic form of resistance against those planning for their future:

> An Italian wanted to start the project of *tiquisque* here. Many of us went to the first meeting. But we soon left because he was just humiliating us. He even said that here are people who lie down and open their mouth in the hope that the food would fall down from the sky. We didn't like such finger-pointing because we are working hard out here. So, we refused to participate any more and the whole project was dropped.

These experiences have provoked a series of protest movements in Río San Juan challenging the amount of funds spent by dozens of development projects with few tangible benefits to local communities. However, these movements are still fragile and their voices are still scattered. This is partly because their loose forms of organisation make these movements invisible, but also because, as newcomers, most of the settlers in Río San Juan have little identification with their locality. The majority of them have come to the region from other parts of Nicaragua after the war, and their

Figure 4 The forest-edge settlement of Las Maravillas, Nicaragua. Photo: Anja
Nygren

life is characterised by mobility and displacement. The social landscape is
composed of multiple actors with diverse backgrounds and ambiguous
intentions: the villages are politically fragmented into Sandinistas versus
Liberals (or ex-Sandinistas versus ex-Contras) and religiously into
Catholics versus Evangelicals. This makes working together on communal
affairs difficult, and in addition, many people are tired of the continuous
drive to organise. According to them, this has been a burden imposed
upon them by developers for decades: During the Sandinista government
they were persuaded to form cooperatives and peasant unions, today they
are urged to form community-based organisations, women's groups and
'friends of the forest' associations.

All this demonstrates how state officials, development experts and
NGOs easily legitimise their projects of resource regulation by creating
images of themselves as protectors of nature and/or facilitators of
development, while at the same time labelling local people as target
groups in need of governance and guidance (for corresponding
inequalities of scale and power, see Anderson and Ellis in this volume).
The state attempts to control the territory of Río San Juan through various conservation programmes, while NGOs direct their attention to
controlling schemes of integrated rural development. For many NGOs,
local empowerment simply means participation, although the issue is
not merely about whether the people can participate, but whether they
have the means to define the terms of their participation (Nederveen

Pieterse 1992). Grassroots partnership is not the same as the local right to control the resource access, which, in the end, is what most of the forest-edge communities in Río San Juan are striving for.

Struggles over resources and environmental images

In the struggles over resources and representations in Río San Juan, diverse images of the environment and local resource-users were used as strategic tools to legitimate conceptions of the 'proper' use of natural resources while invalidating others. This concerned especially the issue of what counts as nature, who controls nature, and what is the relationship of humans to nature: Are they part of it, external to it, or somewhere between?

In the perception of many Nicaraguan conservation authorities, nature was seen as something 'out there', separated from humankind. In this view, any human presence inside Indio-Maíz meant a threat to the reserve's pristine wilderness. Nature was seen as something in opposition to culture, and the tropical forests as a symbol of paradise on the verge of being spoiled by human intervention. This discourse also carried a message of eco-catastrophe; the efforts to protect the Great Reserve of Indio-Maíz were presented as the last chance to save Río San Juan, and even all of humanity, from ecological destruction. This discourse largely followed an international conservationist agenda for tropical conservation, in which the 'last frontier forests' are conceptualised as remnants of pre-human wilderness and their natural inheritance as a common heritage for future generations (see Bryant et al. 1997, Singh 1998).

From this point of view, the local settlers were seen as disruptive forces on the fringes of the protected area: They appeared as reprehensible invaders of a majestic wilderness. The frontier settlements were presented as a 'jungle of eco-disaster' and an 'anarchic forest frontier' in need of control and order. The forest clearing realised by settlers at the edge of the reserve was seen as an outrage against humanity. In an environmental pamphlet provided by a group of conservationists, the contrast was presented as follows:

> Indio-Maíz, I went to know you intimately, to caress your trees, toucans, and parrots. You are the cradle of Nature, teeming with life and mystery … But I also have witnessed how dozens of colonists surge into this region. They come along the narrow road in outworn IFA trucks, packed with hogs, chickens and all kinds of animals. These people rush inward and hack out homesteads everywhere. They are destroying the nature. With migration as high as a hundred people a month, the forest will soon give way to clearing and burning.

The NGOs and development advocates working in Río San Juan generally questioned such an essentialist view of nature as an independent given, outside the human world. By emphasising the role of human beings in establishing what is natural and what counts as nature, they attempted to transcend the strict dualism between 'primeval' and 'human-shaped' landscapes. Many NGOs contested the view of Indio-Maíz as a 'natural habitat irretrievable for science', and attacked the image of the reserve as a 'pristine sanctuary for recreation' as an elitist Western concept. According to them, there is an enormous gap between the Northern 'wilderness' agenda and the Southern 'survival' agenda, as most of the Southern people live at the margins of the environmental discourse of the North.[9] Such was the view of Luís, an active member in a Nicaraguan environmental NGO:

> We think that the only means of conserving the reserve is to use its resources in a sustainable manner. One of the perfect ways would be by extracting certain non-timber products, while the local people would act as guards of the reserve. An ambassador once came here and said that the solution is to place watchtowers with armed guards around the reserve because otherwise the Nicaraguans don't understand. I told him that they were crazy, that when the first watchtower appears, people will meddle ... Indio-Maíz belongs to the category of an absolutely protected reserve, untouchable, untouchable, which is absurd.

In the agendas of these organisations, environmental education was given a high priority. In the workshops organised by rural development advisers to local people, impressive posters were used to contrast the grace and beauty of a standing rainforest with the desolate appearance of the forest-edge colonies. Tree planting and nature protection were invoked as symbols of birth and life, while tree cutting and forest clearing were enshrouded in metaphors of violence and death. According to these advisers, local settlers were imbued with an enormous capacity for conservation, if only correctly inspired. In this context, the ongoing development projects were portrayed as unprecedented opportunities to promote improvements in local environmental morality.

For local settlers, the idea that the forest frontier with its abundant resources could be possessed, 'putting the jungle into production', was still a powerful image. This perception was reinforced by the earlier agrarian legislation, according to which uninhabited forests were 'idle' lands that could be appropriated by clearing the forest. Some older inhabitants of Río San Juan were of the opinion that the state still had vast areas of unoccupied forests, but that it did not want to give them to *campesinos* on account of the current politics of nature protection. The more recent colonists already knew that there was no free land left in Río San Juan; all of them had bought their plots of land at rapidly increasing prices. When asked about the significance of the Indio-Maíz reserve, people first told me about its importance as a 'source of water, pure air to breathe, shade from the blazing sun and protection for poor animals who do not have any place to live due to the barbarous deforestation.' Most of this rhetoric they had heard on the local radio *Voz del Trópico Húmedo*. After repeating this litany, they usually presented another, alternative interpretation of the reserve as a 'reserve of land' (*reserva de la tierra*) to be later distributed for farming to their children and children's children.

The source of this double conception lies in the deeper meanings of forest and nature in the settlers' environmental perceptions. Central to the self-identity of local settlers was their involvement in 'taming the jungle' through hard work. Pioneering was constructed as a project of assiduous people who want to show the fruits of their labour. Many settlers identified themselves as frontier-breakers, gaining mastery over nature by 'making forest yield'. From their perspective, a forest was a symbol of isolation and enclosure; it was an obstacle to be overcome.

In my daily conversations with local inhabitants, they repeatedly expressed a concern for land rights and productive resources. These settlers, who had competed for scarce jobs in their home territory as landless people, had moved to the frontier in search of free land. Now that they had found a piece of land to cultivate, they simply refused to admit that their claim could be insecure or even illegal. These people had experienced many hardships and many inconveniences, being at times close to great dangers and great risks. They accepted the toils and limitations of their lives because they believed that a frontier offered a challenge: a future of much effort and struggle, but also a possibility for building a farm, making a home and raising a family. 'We came to this jungle to brave failure in order to make a home in this wilderness', they often said with considerable pride. By these comments, they wanted to point out that nature is not something to be separated from their social exigencies.

People also told me about the difficulties they encountered in making their home as colonists in a hostile jungle, with jaguars and snakes wandering in their pathways and supernatural beings attacking lonely forest travellers. In their perception, the forest was a symbol of the wildness of nature, which causes rains, storms and other natural hazards and supernatural violence in human communities. Nature was something to be mastered by human forces. This perception was largely misunderstood by conservation authorities, who attributed the settlers' forest-clearing activities to their primordial 'land hunger' or cultural 'forest phobia', with no references to the wider contextual factors – such as agrarian policies, land tenure regimes and market forces – that have reinforced a pattern of forest conversion in Río San Juan for decades. There was also little recognition of these settlers' difficult conditions to meet the basic daily requirements and of their vulnerable positions in relation to a far-reaching global economy.[10]

All this demonstrates how easily authorities impose particular roles and representations upon local people, while at the same time ignoring all the alternative conceptualisations. The dominant representations of the local settlers as 'people with an ardent colonising spirit' have largely shaped the programmes of nature protection and environmental education in Río San Juan. Conservation authorities' images of 'nature destroying local people' legitimised their claims for increasing state authority in the region, while NGOs' view of local people as 'ignorant forest-fellers' justified restrictive development interventions in the people's production systems. In the calendar of a conservation organisation implementing environmental education in Río San Juan, such paternalistic slogans as 'To teach our *campesinos* how to manage their farms is to leave a mark on the Earth forever', exemplify the key role which the developers see for themselves of improving local farmers as the 'real' managers of the productive resources.

Many conservation officials also considered the local resource-users to be fettered by cultural traditions. Many of them constructed local environmental knowledge in such a way as to suggest that, although the local people live in a rich tropical habitat, they are ignorant of how to take care of it: They were deemed to be colonists who know how to tame the jungle with the *machete* but not how to conserve tropical biodiversity[11]. Local settlers responded to these accusations by pointing out that the appeals for local people to change their attitudes towards nature had little relevance when the power to make a significant difference in local resource management was so unequally distributed. They also challenged this display of concern for biodiversity by critically asking who it benefits. According to local inhabitants, the stories of the forests' imminent disappearance and nature's indiscriminate destruction largely come from outsiders who are feeding on sensationalism.

The most critical issue in the visions of the development advocates working in Río San Juan was the search for a simple solution to the region's complex environment-development problems. Each development project entering Río San Juan was presented as a pilot project which would ultimately lead to success. Local settlers challenged this situation by pointing out that in the cycle of different booms and busts, the developers' big promises have rarely been fulfilled. A local extractor, don Ernesto, told me amusingly 'how some *cheles*[12] are going to implement a project of rattan as an alternative non-timber forest product in the community of Buena Vista, although there is almost no rattan left in this region. And all this just because the experts have now realised that the tropical forests are more than timber.' Through this story, don Ernesto wanted to call attention to the ignorance of the developers who had little notion of the wider social and political context in which the utility of local natural resources is continuously defined and redefined.

Conclusion

This chapter presented an account of contested struggles over protection and production which are now common in tropical forests all over the world. Influenced by global concern over tropical deforestation, and by political pressure from international conservation organisations, many developing countries are transforming much of their remaining forests into restricted-use protected areas. These policies have many implications for the livelihood opportunities available to local populations. In the worst case, people may be forcibly evicted from their homes, while in the less violent but no less perilous cases, local inhabitants' possibilities of meeting the demands of everyday life may be severely affected. About 70 percent of the protected areas worldwide are inhabited or regularly used by local people; in Latin America, the figure is about 86 percent (Ghimire and Pimbert 1997: 7). Not surprisingly, conflicts over access to productive resources are common in many of these areas, as carefully demonstrated in many of the chapters of this book.

Nature-based conflicts over tropical forests include multifaceted processes of control and power pursued within complex social relationships. International conservationists, governmental authorities and non-governmental organisations often search to legitimise their control over the 'last tropical forests' by claiming that ultimately they are the only ones who have the necessary knowledge and expertise to devise feasible strategies of protected area management. The resistance of local populations is as much a response to authoritarian policies of conservation and development as to paternalistic forms of institutional governance. An increasing number of people living within or in the fringes of protected areas conceptualised as 'hotspots of global biodiversity' are arguing that local natural resources should form a basis for local production and livelihood, instead of being declared global patrimonies dedicated to strict preservation.

Social categorisations play a crucial role in these struggles over resources and representations. In the perceptions of conservation authorities, tropical forest-dwellers are often portrayed by essentialist images of Indians as 'born naturalists', while settlers and other recent migrants are categorised as 'forest-hostile encroachers' (for more about such images, see Adams, Novellino and Nugent in this volume). These representations affect the general formulation of nature conservation and protected area management policies. The view of local people's nature–society relationships as 'adaptive' or 'destructive' provides justification for paternalist programmes, with little recognition of the fact that the future of nature conservation lies in fruitful cooperation and more equal relations of control and power between conservation authorities and communities surrounding the protected areas. At best, ethnography may provide well-suited methods to elucidate these relations. By their commitment to long-term field studies of local resource-users and their interactions with reserve managers, government officials, non-governmental organisations, national and international conservationists and development advocates, anthropologists may offer valuable analyses of complex processes of nature conservation and social justice.

Acknowledgements

This essay draws on research financed by the Academy of Finland. I am grateful to the people of Río San Juan and to the many ministries, development institutions and NGOs in Nicaragua that cooperated with my field research. Krishna Ghimire, Paul Little and the participants at the 'Ethnographies of Ecological Underprivilege' workshop held on 29th April 2000, at Goldsmiths College, University of London, provided valuable comments on previous versions of this essay. A rather different version of this essay appeared in *Development and Change* 31(4), 2000.

Notes

1   All names are fictitious.
2   For this polemic, see the article by Soulé and Sanjayan (1998), suggesting that the target percentages of the international nature conservation organisations for protecting at least 10 to 12 percent of the total land area in each nation are not sufficient. According to Soulé and Sanjayan (1998), the better estimate to protect most elements of biodiversity is about 50 percent. For criticism of such a view, see Ghimire and Pimbert (1997) and Neumann (1997), according to whom the policy of declaring large areas of forests as protected reserves has a powerful negative impact on the local poor.
3   The reserve originally covered 295,000 hectares of land.
4   Raicilla is a natural medicine used against amoebic dysentery. Both wild and cultivated raicilla were extracted and exported from Río San Juan to Europe and the United States, where a commercial medicine was processed from its subproduct. In the late 1970s the economy of raicilla suddenly declined in Río San Juan. The principal cause was the rapid depletion of the natural supplies of raicilla due to over-exploitation (Offen 1992, Didier 1993).
5   Of course, it would be unfair to claim that all conservation authorities had such a monolithic view of protected area management and that all of them have been insensitive to the needs of local people. The conservationists often presented the matters wrapped in the conventional rhetoric during the first interviews; they spoke with self-assured tones about their projects as 'pilot projects of successful conservation'. In daily conversations, their comments were more tentative and they showed more awareness of the complexity of the situation.
6   For detailed analyses on the role of NGOs in current development policies, see Vivian (1994), Edwards and Hulme (1996), Bebbington (1997), and Fisher (1997).
7   I am well aware that categories such as 'NGOs' can also be fairly essentialist. In reality, the NGOs represent a vast range of different types of organisations, with different set of personnel and different backgrounds and histories. For more in this respect, see Hadenius and Uggla (1996).

8   For criticism of the ignorance of local communities' internal differentiation in development politics, see Pigg (1992), Cleary (1993) and Agarwal (1997).

9   For discussion on the 'environmentalism of the rich' versus 'environmentalism of the poor', see Gadgil and Guha (1994), Cronon (1996), and Guha and Martínez-Alier (1997).

10 For social representations of non-indigenous peasants in Amazonia, see the studies by Nugent (1993) and Harris (1998). For the picture painted by development experts of local people and environmental histories in various parts of Africa, see the studies by Fairhead and Leach (1995), Rocheleau et al. (1995), and Neumann (1998).

11 For further discussion of struggles over different knowledge systems, see Nygren (1999).

12 A pejorative appellation, referring to a person who is light-complexioned and foreign (North American or European).

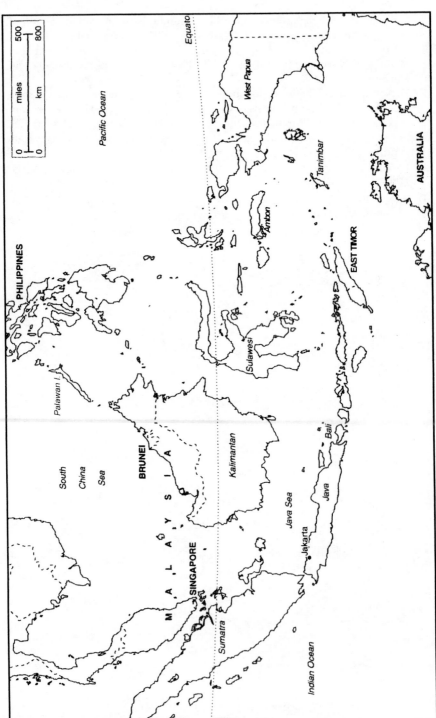

Figure 5 Indonesia and The Philippines, showing Tanimbar Island and Palawan Island

3

THE ENVIRONMENT AT THE PERIPHERY
Conflicting Discourses on the Forest in Tanimbar, Eastern Indonesia

Nicola Frost and Rachel Wrangham

The Tanimbar Islands lie in Eastern Indonesia, far from regional and provincial centres. The few anthropological studies which have been carried out are largely structuralist accounts of kinship relations (e.g., Pauwels 1985, McKinnon 1991), and the islands' apparent impenetrability is further enhanced by their mysterious local reputation. However, invisible as they may be in regional and anthropological accounts, in recent years they have gained an international profile as a result of the strength and endurance of reactions to an unwelcome logging concession.

As related on a number of Internet sites the story is dramatic (see, for example, Earthaction website 1997). Following the unexpected and unannounced arrival of a private logging concession on the island of Yamdena in 1991,[1] a group of local people marched on the logging camp. In their protest they damaged machinery, and in the ensuing fracas, which apparently involved the local military, one villager was shot in the leg. In the aftermath many people, including village leaders, were given jail sentences. Although this concession was subsequently revoked, in 1995 a new concession was granted to a branch of the Indonesian parastatal, Inhutani. The websites invoke a variety of international advocacy agendas, and stress the need for continuing international support of a campaign of legal challenges and ministry-level lobbying being coordinated by ICTI (*Ikatan Cendekiawan Tanimbar Indonesia*, the Tanimbarese Intellectuals Association), a Jakarta-based group of Tanimbarese.

Arriving in Jakarta in March 1998, just after Suharto had been re-elected as President and before his dramatic fall from office, we talked to some of the members of ICTI about their attempts to dislodge the concession. Coming ourselves from a background of advocacy for indigenous rights we were impressed by the organisation of their campaign, and their persistence against all odds. However, once in Tanimbar it became clear that the narrative of ICTI was just one element in a much more complex and dynamic set of discourses about the Tanimbarese forest. In this chapter we examine the debates which have coalesced around the concessions, and the widely varying priorities and perceptions of the different groups and interests involved. We examine the framework of national–local power relations with respect to natural resource management, and explore the deep conflicts which exist within this framework. We will also aim to demonstrate the value of the ethnographic approach in unpacking the layers of power and interest visible here.

Peripheral islands?

The Tanimbar Islands lie at the southern edge of the province of Maluku, Eastern Indonesia (Figure 5). They are marginalised – most simply because of their geographical remoteness, but also as a result of a range of historical and cultural factors. At the same time, owing to the nature of the centralising Indonesian state, they remain tightly bound into the political and economic life of the nation. An understanding of this complex and power-laden relationship between centre and periphery, between national government and villager, is crucial for our discussions of the ways in which different discourses intersect and interact as it shapes the location in which these interactions can occur. This section outlines some of the factors that shape everyday life in Tanimbar, emphasising the isolation of the islands, and looks at the ways in which central government has circumscribed the agency of local actors.

It is hard to overemphasise the geographical and administrative isolation of the Tanimbar Islands (see Figure 5). It takes most people at least four days to travel to the provincial capital of Ambon, even though the distance is only 400 miles. In the past, a regular plane service allowed wealthy people and time-pressed tourists to reach Ambon in just over an hour, but the economic crisis meant that this service had all but ceased during our fieldwork period. Yet, though the connection with Ambon is crucially important for all administrative matters, unlike other Malukan islands Tanimbar's economic links are largely to the west: a passenger ship calls each month from Java, and cargo ships owned by wealthy Chinese merchants make regular journeys to Surabaya. Within the Tanimbarese archipelago as well, travel and communication is difficult. The few existing roads are frequently rendered impassable by poor maintenance and bad weather, while sea travel is only possible for six months of the year. Villagers and schoolchildren from the western coast often have to cross the island to Saumlaki on foot, a journey of over forty kilometres through the forest. Electricity is severely limited, with supplies largely confined to the main towns and neighbouring villages, while there are telephones only in Saumlaki and its immediate surroundings.

However, it is not simply for geographical reasons that Tanimbar is isolated: its historical and cultural trajectory has set it apart from other islands in Maluku, and the Tanimbarese are still regarded with some suspicion by people from neighbouring islands. The greatest difference is that whilst much of Maluku was colonised by the Dutch in the very early years of their East Indian Empire (as they pursued their spice-trading monopoly), the ferocity and poverty of the Tanimbarese meant that Tanimbar was not 'pacified' by the Dutch until the early twentieth century. The reputation for ferocity and black magic is still alive, though much attenuated, with the Tanimbarese and those from the islands further to the west still seen as strange and different (see, for example, Forbes 1987, 1989).

Tanimbar is not merely geographically and culturally isolated: it is peripheral politically. The Indonesian state under the New Order was a centralised and centralising state, which has taken active steps since the 1960s to reduce the power of local leaders. This marginalisation is by no means specific to Tanimbar as it is largely the result of a series of laws on village government discussed below. However, Tanimbar's cultural and geographical distance from Ambon, the provincial capital, puts it in a particularly poor position regarding funding decisions, implementation of policies and quality of personnel. Moreover, Tanimbar is administratively grouped with numerous other geographically, topologically and culturally distant and disparate islands, which further reduces the potential for local involvement in decision-making. During the period of our research ICTI was resisting the proposal that the Tanimbar archipelago should be split between two administrative regions, which would have even further reduced local agency. During the New Order, the region of Southeast Maluku was at the receiving end of centralised legislative decisions and top-down development schemes, and had no opportunity to change these schemes or make other suggestions. This problem was particularly acute with regard to the use and management of natural resources, and is indicative of the way in which the centralised Indonesian state managed its relations with villages, particularly those in the outer islands.

Administrative marginalisation goes much further than mere lack of consultation, however destructive that is. In common with all Indonesian villages, those in Tanimbar have suffered greatly as a result of Indonesia's laws on village government enacted in the late 1970s. These laws drew on an institutionalised suspicion of traditional *adat* leadership, the desire to entrench the power of the ruling Golkar Party, and the need to have a local level bureaucracy capable of implementing government regulations. Their effect has been to centralise an enormous amount of power in the hands of the *kepala desa* (village head), so that without a *kepala desa* the village is paralysed: no subsidies, no development funds, no women's groups, and problems with health and education provision. In Tanimbar these problems were vastly complicated by the partial continuation of the parallel *adat* system of leadership and government (for comparative examples see Warren 1993, Antlov 1995, Spyer 1996).

Prior to the implementation of these laws on village government Tanimbarese villages were organised around hereditary leaders. Each clan had a position within government, which was conceptualised as a boat: one family took the position of captain, another as steersman, while others rowed or kept watch in the prow. A certain amount of flexibility was possible, so that the leader of the village could come from any of the families said to possess a *batu adat* (literally, a rock of tradition, meaning the right to rule). The 1979 law altered all this when it directed that any adult, providing they fulfilled certain legal requirements, could become the *kepala desa*. These requirements include a basic level of education, a religion and Indonesian citizenship. The effect of the new system, operating on top of the old, was to almost put an end to local leadership and initiative. Disagreements lasting many years resulted from the nomination of multiple candidates from different village factions, and the regency-level bureaucracy has contributed to local paralysis by taking years over its part in decision-making. Villages were destabilised, and in many cases no local-level funding has been received (or at least accounted for) for many years. In a state where central control has been so important for so long, lacking access to funds and information coming from the centre has had a pernicious effect. A lack of local leadership and local legitimacy has also had a harmful effect on local resource management and local responses to outside interventions. Systems of resource management – which rely on respect for local authority rather than coercion – have broken down in many areas, and coherent collective action in response to external claims on resources (such as the anti-logging protests) often becomes subordinate to intra-village politics.

Given this heritage of disempowerment and marginalisation, it is unclear whether new legislation concerning local government will be successful in reversing these effects.

Tanimbar's forests: Transformations in resource use

Land use, and hence forest use, in Tanimbar has undergone significant changes over the last two or three decades (Sunderlin and Resosudarmo 1996). Some of these relate to the logging concession, while others are related to the changing world prices of agricultural commodities, improved transport links and the islands' increasing population. In this section we will look briefly at how agriculture in Tanimbar currently operates and at the changes and transformations it has undergone.

It is still the case that nearly all households in Tanimbar continue to farm, or are reliant on close relatives who do. They practise swidden agriculture. Most household food is grown in gardens lying at a distance from the house, with the major crops consisting of tubers such as taro and sweet potato, some dryland rice varieties and a few vegetables. New gardens are cleared every year or so as the topsoil is very thin and loses fertility quickly. Much of this clearance occurs in primary forest, which has become more accessible as transport has improved. As it has become more profitable to grow and sell copra (a coconut product), the rate of forest clearance has increased markedly. Copra has generated cash, which has been used in the first place for school fees, basic commodities such as soap, tea, salt and imported rice, then for concrete and corrugated zinc to build permanent houses, and increasingly for electrical appliances, satellite dishes and chain-saws. A process of differentiation both within and (so far) more importantly between villages is clearly underway, with some villages making large material gains as a result of selling wild buffalo or valuable marine products.

With the exception of the numerous civil servants based in Saumlaki, the income of those who do not rely directly on farming is generally still based on agriculture. The Chinese[2] merchants dominate this group of intermediaries, buying all cash crops within Tanimbar (copra, marine products, orchids, parrots and, in the past, crocodile skins) and transporting them on to Java for sale. Their linkages to networks in Java provide crucial access to wider markets, but their fortunes are clearly linked to the profitability of Tanimbarese agriculture and access to the forest.

At the time of our fieldwork in early 1998 the Southeast Asian economic crisis was really beginning to have an impact in Tanimbar. Prices of many household goods rose sharply or the goods became unavailable. However, at the same time prices paid for Tanimbarese export crops, in particular copra and marine products, rose dramatically.[3] This increased the incentive for local people to concentrate on their coconut groves, and to strip as much marine produce as possible from the sea. This was exacerbated by the climatic effects of El Niño and La Niña which, whilst not as severe as in some areas of Southeast Asia, meant that the annual crop cycle was disrupted, leading to a prolonged 'hungry season' and the need for money to buy food. This may well have been a temporary change: from our short study it is impossible to tell. However it fits with an increased focus on short-term cash gains, rather than on sustainable resource management, and may also represent an intensification of a general trend whereby cash crops were becoming more of a central feature of many households' economies, rather than an occasional supplement.

Yet so far a combination of poor transport, relatively low population and limited markets have meant that Tanimbar's forests have been uniquely preserved within Maluku. Tanimbar has the only remaining primary forest in southern Maluku, which harbours a number of rare bird and tree species. Forest land is used not only for swidden farming but also for wood for building, fuelwood, for hunting (pigs and buffalo), trapping birds for sale and gathering fruits. However, in 1991, despite these long-standing local practices, which are locally reputed to have been recognised by the Dutch, and despite an earlier decree giving protection to the island, central government granted a logging concession to the privately owned company PT Alam Nusa Segar, which extended over most of the forest. There was no local consultation and, as a result, there was considerable fear and indignation, resulting in the violent confrontation described above. Around the same time, the provincial government began to enforce a change in local government legislation, which limited village heads to two terms in office, and several long-standing *kepala desa* were forced to stand down.

PT Inhutani I, the parastatal which took over the scaled-down concession in 1995, is responsible not only for the 'management' (*pengelolaan*)[4] of the forest, but also for carrying out a range of social development and reforestation functions. Their job has not been easy, and they state that they are continuing to lose considerable amounts of money in Tanimbar. There are a number of reasons for this. One has been a lack of knowledge about *kayu torem*, a hardwood said to be unique to Tanimbar – this has meant that it has been hard to find a market for the wood. Inhutani was also at first unaware that it is necessary to saw *kayu torem* while it is still wet, with the result that vast piles of logs, many split and irreparably damaged by the elements, now lie abandoned at the quayside. Moreover, Tanimbar is a difficult place in which to work: transport costs are high, local people are only employed for unskilled work and the weather prevents logging for more than half the year. Nor are the rewards high: Inhutani staff state that the maximum permitted potential yield (based on the species and size of tree that may be cut) is only two to three trees per hectare, far below that of Kalimantan and elsewhere. In addition, Inhutani must, in theory at least, abide by the extensive list of regulations on replanting and regeneration of degraded forest land (Inhutani 1 website 2000). As a result of all these problems, only a small part of the total extent of the concession, largely restricted to areas in the south within reach of logging roads built by PT Alam Nusa Segar, had been harvested by 1998. Yet despite these obstacles, Inhutani remains in Tanimbar.

Discourses on environmental change and development

As the powerful reactions to the granting and operation of the logging concession indicate, the environment is a highly charged issue in Tanimbar, and the forest is a powerful focus of discussion for many groups concerned with Tanimbar. In the following sections we identify three different discourses which intersect in the forests of Tanimbar. All imbue the forest with great symbolic importance, but as we will indicate, these discourses concern themselves with very different issues. It is in the forest that they come together and in the forest that discussions between groups with very different power bases are able to occur.

Most important is the discourse of Indonesian national development which underpins all government and parastatal activity. In Tanimbar the discourse of national development is most salient regarding the forest, as language deriving directly from the Indonesian Constitution is used to define its forests as part of the divinely conferred national resource base, which must be used to further national development goals. A second discourse about the forest reacts directly to this, arguing that the question of the forest is a question of local heritage and of conservation: this shifts the agenda to one of rights, duties and values ascribed to knowledge. In Tanimbar this is associated with ICTI, and through them with the international campaigning organisations that support them. For most local farmers the question of the forest is more a question of land and livelihood security, and the continued opportunity to develop personal wealth without restriction, through the cultivation of cash crops and a continuing pattern of annual forest clearance for swidden agriculture.

It is crucial to remember throughout the next sections the imbalance of power that we have already described which exists between different actors and these different discourses. Though we give each of the discourses approximately equal weight, they speak at totally different volumes: the discourse of national development is the one that effectively dictates and circumscribes the areas which can be discussed, through absorbing into its own agenda elements from the other two.

Pembangunan: The imperative of national development

It is in the area of forests that the discourse of Indonesian national development impinges most immediately on Tanimbar. The discourse itself is all-embracing, a national mantra of development which informs all government development initiatives, ranging from agricultural extension and village subsidies to golf courses and transmigration policy. The need for national development and the state's duty to use national natural resources to achieve it can be traced back to the Constitution of 1945. Chapter 33 states clearly that it is the right and responsibility of the state to control (*menguasai*) all natural resources for the general good of the Indonesian people.[5] This has been the central tenet of forest 'management' under the New Order, and it became the basis of the 1967 Forestry Law, which until 1999 regulated all forest-related activities (for more details see Wrangham, 2002).[6] It is a tenet that government has been fearful of relinquishing; as such a statement of the central right to control and 'manage' is also a powerful legitimisation for the nature and character, even the existence, of the Indonesian state. It is with reference to this right that the New Order state awarded thousands of logging concessions to commercial companies from the late 1960s onwards. No pre-existing rights were recognised, no compensation was given, and other than reference back to the Constitution, no justification was considered necessary.

For the state, then, the forest is a potent symbol of Indonesia's wealth of natural resources, and the key to its continued development as a unified, heavily centralised nation. However the forest also conceals potentially 'uncivilised', undesirable social groups, which impede these national development aims. Forested areas tend to be remote, inaccessible places, often populated with people who deviate from the image privileged by the Java-based government, of settled, rice-growing intensive agriculturalists. These people – nomadic hunter-gatherers or shifting cultivators – are designated *suku terasing* (isolated people), in need of forcible settlement and civilisation to transform them into proper citizens.[7] Although not officially *suku terasing*, to an extent the Tanimbarese fall into this category through their physical and cultural remoteness from Javanese norms (see Li 1999: 18). In this respect, the state's 'forest' is at once a firmly depopulated, controlled space for national development, and a source of threat and insecurity far from the centre of power (cf. Scott 1998).

In the Tanimbarese context, Inhutani is a symbol and agent of this centralised control. The language of the Constitution informs Inhutani's rhetoric when describing its role in Tanimbar. The objectives on Inhutani's (English-language) website see forestry as merely an instrumental means by which to further the overarching aim of national development:

> To take active participation in the implementation of, and to provide support for, the Government Policy and Program in economics field and the national development in general and forestry sector development in particular. (Inhutani1 website 2000)

Inhutani staff speak of their tripartite 'mission': environmental conservation and management, social development and commercial exploitation. The aims of preserving trees and developing people are presented and acted on in very similar ways, demonstrating the clear conceptual connection between forests and national development as part of a single, complementary vision. Inhutani speaks of developing local people through the use of the forest, and emphasises the social side of their affairs. According to national development discourse, problems and obstacles in the operation of programmes like reforestation and community forests are due to the ignorance of the local people who are meant to benefit: they are not culpable, merely uninformed. The discourse allows no space for error or adjustment on the part of the developers, or for conscious resistance or sabotage on that of the developed.[8]

As a parastatal with responsibility for regional development, Inhutani implements a number of social development programmes. These include a soft-loan scheme, student scholarships and funding for school libraries, new health clinics and local road improvement. While none of these projects has met with much local opposition, there have been ongoing problems with the reforestation schemes that Inhutani has sponsored. These schemes, which involve replanting wasteland with a mixture of trees to provide fruit and timber, have been plagued by fires, which may well have been started deliberately. Whatever the case, it is certain that Inhutani will not allow the failure of the reforestation programme to weaken the power of its development mandate.

Thus it is the case that the forest is an important part of this discourse of national development. But the discourse of national development is also crucial to everyone else's understanding of what the forest is and what can be done there. This is the discourse which therefore can be seen as defining the boundaries within which others can or cannot speak. In most cases the result is silence: the power difference between the state and the villager is just too great. In Tanimbar it has only been as a result of the mediating intervention of the Tanimbarese Intellectuals Association (ICTI), themselves drawing on internationally valid discourses of rights and conservation, that conversation has been possible. Thus while the discourse of national development frames the general setting within which the forest can be discussed, the rhetoric employed by ICTI, and absorbed to some degree into this national discourse, also shapes the situation.

Connections with global discourses – ICTI

In this section we look at the complicated and shifting alliances that ICTI have built up and built on as part of their campaign to rid Tanimbar of logging concessions, and at the language they have used in so doing. Through the creation of these alliances they have developed two distinct and complex roles: on the one hand, representatives of Tanimbar to the outside world, and on the other, mediators between 'the Tanimbarese' and the government. In a lesser sense they have also had to mediate between the goals of the international organisations who support them, and those of the people whom they see themselves representing.

ICTI carries out its campaigning work in the name of the people of Tanimbar. It claims that anyone from Tanimbar is automatically a 'member' of ICTI. Thus from their point of view ICTI members in Jakarta are not *representing* the Tanimbarese through explicit mandate, rather they *are* 'the Tanimbarese'. But because of their role in mediating between and facilitating interactions between Tanimbarese farmers and the government we would argue that the position is not that simple. As a result of their distance from Tanimbarese everyday life, their exposure to different ways of thinking in their lives in Jakarta, and their involvement in international campaigning movements, the way they portray 'the Tanimbarese' and the points that they stress as important in Tanimbarese relations to the forest are different from those emphasised by Tanimbarese farmers.

There has been a historical progression in the elaboration of ICTI's argument about the rights and wrongs of logging in Tanimbar which parallels both the change in international environmental discourses and changes within the Indonesian context itself. The first broad campaign that ICTI became associated with was conservationism, which advocates preserving areas of environmental significance and protecting them from overexploitation. In this area they have worked closely with the environmental advocacy network EarthAction, the Indonesian forest conservation NGO, SKEPHI, and Down to Earth, a UK-based lobbying organisation devoted to ecological justice in Indonesia.

Documents resulting from this engagement emphasise the biological uniqueness of Tanimbar's fauna and flora (see, for example, ICTI 1993). They emphasise the delicate ecological balance that maintains the island's ecosystem: without the forest cover they argue that the thin topsoil will be quickly eroded, leaving a barren limestone outcrop, which itself could quickly be eroded to nothing. In some cases dramatic statements have been made about the danger of the island disappearing altogether (see, for example, Nativenet website 2000). In line with a stringently conservationist view, ICTI members reject any suggestion that any level of logging or timber management can be sustained by the Tanimbarese environment, and categorically state that the results of the environmental impact surveys carried out by scientists from Pattimura University in Ambon before the Inhutani concession was granted were 'fixed' to show that controlled logging would not harm the island.

Although there is no clear chronology and ICTI's engagement with the conservation agenda has continued, during the 1990s it became more closely involved with a second international campaign, that for the rights of indigenous peoples. ICTI consciously characterise the inhabitants of Tanimbar as 'indigenous people'. On one level this is a reasonably accurate description since most people who live in Tanimbar were born there, have family there, and the population is relatively culturally and ethnically homogenous. However, as part of international environmental activist vocabulary, the phrase carries with it strong images and assumptions. The image is generally of a remote, isolated band of people, probably hunter-gatherers or shifting cultivators, who are ethnically and socially homogenous, and relatively self-sufficient, not participating in the market economy, or in formal education, much less in the national community. Indigenous peoples are seen as unchanging, more connected with nature than history; they are the 'guardians of the rainforest', possessed of a special wisdom regarding the use and management of natural resources. As in the situations described in other chapters in this book, by no means all of this applies to people in Tanimbar, or at least not in a straightforward way. However, as Gupta (1998) argues, the rhetorical strength of the 'indigenous' identity, combining as it does a sense of vulnerability with a celebration of the 'noble savage', privileges it over other potential identifying characteristics (for example, poverty), particularly in seeking engagement with Western agendas.

There are many reasons why ICTI has shifted the emphasis of its campaign away from conservation and towards indigenous people. One is the shift in international discourses, which has led to more attention being paid to human rights than biodiversity. The alacrity with which several Western groups picked up the case demonstrates how familiar and readily communicable the story of powerless (and blameless) primitives struggling against the environmental ravages of capitalism has become. Survival International has been involved in the Tanimbar case, as has the International Alliance for Indigenous People.[9] Another reason why the indigenous rights agenda has become more important has to do with the Indonesian political context. ICTI has operated for most of its life under one of Southeast Asia's more repressive regimes. Survival International points out that many activist groups operating under these conditions identify themselves as environmental campaigners as the environment is often seen as less politically contentious, and they are thus more likely to be allowed to continue operating (Sophie Grigg, pers. com. September 1998). This would certainly appear to have been the case in Indonesia where the conservation discourse, although closely linked to issues of power and equity, is not commonly seen as a political topic, whereas the concept of 'indigenous people' is inextricably linked with a radical anti-state agenda.

The 'forest' for ICTI is therefore a complex and contingent symbol, at once expressing a radical agenda advocating indigenous rights and, at the same time, signalling a conservationist stance, which stresses the importance of maintaining the forest as a pristine, protected zone, inaccessible to loggers and farmers alike. For ICTI the forest is central in a way it is not for the other two discourses we discuss here: it is the issue on which the entire existence and mandate of ICTI stands. That said, at a conceptual level it is vital to remember that ICTI do not control the substance of the argument, and that the debate operates *within* the context of Indonesian national development, despite the somewhat broadening impact of international discourses.

Perekonomian Rakyat: Local livelihoods

In stark contrast to Inhutani staff and ICTI activists, when people in Tanimbar talk about the environment and the forest, they generally speak in terms of concerns for livelihoods – 'forests' for local people very clearly contain people as well as trees. The grounds for their concern are clear: they feel increasingly squeezed by rising population and land restrictions, some of which result from the logging concession. They increasingly feel the need for a cash income, so the incentives to plant coconut palms to produce copra increase each year. It is important to note, however, that pressure on land is only just starting to have practical effects – in the villages we visited, it was not clear that actual practices had yet altered very much. However, there is uncertainty about the future, which could be contributing to the apparent increase in 'unsustainable' land practices.

Each village lays claim to an area surrounding the settlement, extending deep into the forest, so that the entire island is shared between different villages. Although certain remote villages have access to far more land than villages closer to the capital, this is tempered by the increased costs of transporting cash crops to market. As more and more land close to the village is covered in mature and productive coconut groves, people are forced further into the forest to clear new land for agriculture, and to expand their plantations. Any fallowing system that may have been practised in the past is now in abeyance, as villagers aim to maximise the size of their gardens, partly in the fear that village lands (now bounded by the logging concession) will shortly be used up.[10]

The fear of land shortages was most intense in villages along the eastern coast, where the population is denser, and the presence of Inhutani most visible. However, most local farmers have had little direct contact with Inhutani officials and activities. In 1998 logging had only just begun to restrict the areas where they could open new gardens, so most opinions expressed were referred to potential and perceived threats. So far experience has shown that Inhutani is fairly tolerant of farmers' small-scale incursions into the concession area, at least for the present. But farmers are worried about the future availability of land for their children and grandchildren (cf. Nygren, this volume). There are also concerns about the forest itself, used as a hunting ground for wild pigs and buffalo, and a limited source for other non-timber products such as sago, roofing materials and even orchids. One village leader also expressed concerns about the future of the village water supply. All along the eastern coast the feeling of being constrained to 'narrow' (*sempit*) land has increased the intensity of boundary disputes between villages, which regularly spark into vicious fights.

As yet local people have faced few real restrictions on their livelihoods, though the perceived threat to them is growing. Fears over wood shortages are starting to emerge, with rumours about restrictions on wood to be used in construction. However, this has not spurred local attempts at reforestation, even in a village where the large number of wood carvers rely on fast-diminishing ebony to make statues to sell to the handful of visiting tourists, and in Ambon and Jakarta. So far very little wood is sold by local people, and trees are not, of themselves, seen as a means of livelihood. There are considerable legal and logistical restrictions on local sale; as there is obviously no local market for the wood, people come up against the Chinese monopoly on cargo shipping. In addition, a licence is required from the forestry department for extraction of wood for commercial purposes. We learned of one village-based project that had obtained a forestry department permit, and was planning a scheme by which timber frames would be built for houses for poor families, and the cement and zinc would be supplied from the proceeds of small-scale logging. It is possible that this kind of project will become more feasible now that more people own or have access to chainsaws. There were indications that some wood may have been cut by local people in the hope of sale to Inhutani, but we can by no means be sure about this.

It appears that local concerns about the forest are less concerns about trees, conservation, or their physical environment, but rather that the forest is used to express deeper concerns about security and livelihoods. Most do not consider trees as a commodity with cash value, nor do they wish to preserve the forest in a pristine state. Local people's priorities when it comes to the forest are therefore safeguarding their livelihoods, both at subsistence level, and in terms of their ability to participate in the cash economy.

Yet although certain government and Inhutani officials would suggest otherwise, local relationships with the forest are not purely utilitarian (for another example of this, see Nygren, this volume). Although villages are now located solely along the coast, in the past they were located deep within the forest, often on higher ground, and were heavily fortified as the result of the considerable inter-village conflict. The move to the coast took place after the Dutch took control of the islands, ended the inter-village warfare, and decided that coastal settlements would be easier to oversee. However, the forest retains considerable spiritual significance. Although nearly all Tanimbarese people are either Catholic or Protestant, there remains a strong parallel system of animist beliefs and practices based on the authority of the ancestors. Before missionaries confiscated them, wooden statues of ancestors were placed in the forest surrounding the village as guardians, and villagers speak of particular trees and stones which mark the inter-village borders, deep in the forest. Even today, permission is needed from the ancestors before outsiders can enter the forest, even if this permission is not automatically sought. We experienced this at first hand when being taken to the site of a disused gold mine: after two hours struggling through the undergrowth, our guides announced they were lost. When we talked about the incident afterwards it became clear that the guides believed that the ancestors had hidden the way because we had not had permission to enter the forest. Thus although the significance of the forest to local people has changed apace with historical and economic shifts, it would be inaccurate to suggest that the Tanimbarese approach to the forest is based solely on utilitarian principles. Rather, Escobar's phrase 'meanings/uses' (1999) is helpful here to describe the relationship between local historical and cultural processes and broader contexts of power in generating local cultural models of nature.

Yet the most fundamental point to recognise here is a lack of agency, which is firmly felt at the local level. The reality of their powerlessness was brought home to the Tanimbarese in the aftermath of the futile direct action of 1992, which left several injured and many imprisoned. Thus though anti-logging feeling has not diminished, people are now understandably far more circumspect about open condemnation – Inhutani is, after all, a state organ, and action against it is unwise and almost certainly futile. There are no opportunities for local participation in decision-making about the management of Tanimbar's forests and other natural resources, and the overall sense is one of suspicion, powerlessness and confusion.

Environmental underprivilege, or privileging the environment?

The three different perspectives on the forest that we have outlined above are not mutually exclusive. That said, representatives of the various perspectives can interpret the same event in different ways. An example of this is the *sirih pinang* ceremony which was performed when Inhutani formally took over the forest concession. This ceremony is a local *adat* ritual in which permission is granted for the use of village land. In a brochure produced nationally by Inhutani, it is stated that Inhutani was invited to take over the management of the Tanimbar forest by local people themselves (through local government channels), in order to prevent illegal logging and to boost economic development. The Tanimbarese then reportedly expressed their support for Inhutani through the *sirih pinang* ceremony, and through making the director of the company a *tua adat* (customary leader). In contrast to this, local Inhutani staff use Inhutani's participation in the *sirih pinang* ceremony as evidence of their sensitivity to local customs, compared with the previous, private logging company and stress how proactive they were in seeking permission in this way. In a third interpretation ICTI dismiss the whole procedure as ill-disguised bribery, with no relation to true *adat* practices.

Individual people can, however, shift between discourses as they move between different contexts. Thus village leaders will express one set of opinions related to national development goals in official village meetings, while expressing their concerns about the future of their livelihoods when talking privately at home. It is important to note that the alliances formed through this constantly shifting network are often contingent, 'single-issue' coalitions rather than the foundations of more extensive communities of interest. For example, long-standing boundary feuds did not prevent collective action against the loggers, but nor was a long-term feeling of solidarity established through this joint action.

Postscript

The situation in Tanimbar has changed significantly since our fieldwork ended in August 1998. Since January 1999 many parts of Maluku have been rocked by ongoing violence between Christians and Muslims, which has resulted in a death toll to date of at least 5000 people. The causes of the conflict are linked to continuing national instability, long-running Christian concerns about immigration of Muslims into the province and fears about Muslim domination of provincial government. Although Tanimbar, with an almost exclusively Christian population, has escaped the worst of the violence, the ongoing climate of instability has affected shipping routes, communications and supplies, as well as having an impact on individual families.

Fundamental changes have been occurring at the root of the Indonesian state. In 1999 new laws allowing for considerable decentralisation of administrative functions replaced the laws on village government. Though confused, and in many parts contradictory, these new laws may provide for far greater local autonomy, and for greater accomodation of traditional systems of leadership and government. Nevertheless, natural resources remain the focus of a tug-of-war between central and local authorities. A new Forestry Law was also passed in 1999 which may have significant effects on forest management at the local level.

ICTI's work has also changed in post-Suharto Indonesia. Their efforts to stop the logging have been intermittently successful, as the different levels of bureaucracy have disagreed over whether or not permits should be revoked. ICTI have also changed their rhetoric, fitting in with the new Indonesia-wide emphasis on *adat* rights and rights to land. Inhutani has since left Tanimbar, though the concession has yet to be officially revoked. The forest has thus remained central to Tanimbarese concerns, though the form of those concerns has shifted slightly amidst the political and economic changes of the last two years.

Acknowledgements

The paper draws on field research conducted in south Tanimbar, Indonesia, from April to August 1998. We would like to thank Sabtu Ohoirat, our research assistant, for his enormous and invaluable help in all aspects of our fieldwork. We would also like to acknowledge the financial support of the Royal Geographical Society and the William Wyatt Fund of Gonville and Caius College, Cambridge.

Notes

1   Our fieldwork focused on the southern half of Yamdena, the largest island of the Tanimbar archipelago, which is where the logging concession is, and where most of Tanimbar's forests are found. However, in common with those we talked to, we refer to 'Tanimbar' and 'Tanimbarese' throughout this chapter.
2   Throughout this article we use the word 'Chinese' as it is used in Tanimbar. The group referred to is internally cohesive, is involved in trade, has important family links with other 'Chinese' in Java, and tends to intermarry. The 'Chinese' have been in Tanimbar for about one hundred years.
3   For more details of the impact of the crisis on forest dwellers see Sunderlin 1999, Sunderlin et al. 2000.
4   *Pengelolaan* is a term that implies commercial forest exploitation; it is thus a concept that is disputed by others involved in Tanimbar's forests, and does not have the moral neutrality of the English word management.
5   Compare this with a similar situation in Syria, described in Chatty, this volume.
6   It should be noted, though the point is not pursued here, that the Indonesian Constitution is in fact ambivalent about the rights of the state: Chapter 18 recognises the legitimacy of traditional rights and traditional legal systems.
7   See Nygren, this volume, for a discussion of a similar 'civilising' role of the state in Nicaragua. There is also an extensive literature on this issue in other parts of Indonesia, see in particular Peluso 1992, Tsing 1993, Li 1999.
8   It is now claimed that 'the label "obstructer of development" has replaced "communist" as the accusation of choice for allegedly subversive activity' (McCully 1996: 264).
9   ICTI has particularly strong links with the International Alliance, being its representative in the 'Bahasa' region (Indonesia and Malaysia).
10 We did not gain any clear sense of how farming practices have changed over the years.

PART II

DISTRIBUTING JUSTICE WITHIN PROTECTED LANDSCAPES

4

PROTEST, CONFLICT AND LITIGATION
Dissent or Libel in Resistance to a Conservancy in North-west Namibia

Sian Sullivan

Introduction

In the last fifteen years, environmental anthropologists, environmental historians, and political ecologists have tended towards a 'corrective and anti-colonial' narrative (Beinart 2000: 270), drawing attention to the ways that environmental discourses can extend a 'northern' hegemony over independent states via donor-funded environment and development programmes (for the African context see Homewood and Rodgers 1987, Fairhead and Leach 1996, Leach and Mearns 1996). Escobar (1996: 56) argues that concepts such as 'sustainable development', 'degradation', and 'community' imply a 'semiotic conquest of social life by expert discourses and economistic conceptions'. The constructed and contingent nature of these concepts and goals has been emphasised by several authors, focussing, for example, on the absence of biophysical 'evidence' for degradation narratives and on the occlusion of local descriptions of, and explanations for, environmental change (Fairhead and Leach 1996). Brosius et al. (1998: 159) argue that an unfortunate outcome of such analyses frequently has been an increasing divergence between advocacy in environment and development on the one hand and, on the other, a more abstract academic critique of the concepts on which many environmentalist movements build, including community, rights, management, resources and degradation. They suggest, however, that there is a critical need for case histories which reflect on and examine the development, applications and consequences of such concepts. An ethnographic approach that details the unfolding of local–global interpenetrations in the environmental arena can provide public space for a variety of views, including those of local individuals. In this way, ethnography might play a significant role in introducing the richness and complexity of real experiences into environmental policy debates.

In the interrelated arenas of environment and development, the Namibian context is dominated by two concerns: first, that land productivity is deteriorating through the overuse and damage of biophysical resources. This is framed primarily as 'desertification' due to overgrazing by livestock and is championed by a mainly GTZ-funded[1] Programme to Combat Desertification (Seely and Jacobson 1994, Wolters 1994, Seely et al. 1995: 57–61, Dewdney 1996, Mouton et al. 1997). The second related concern is for the effects of human activities on biodiversity, manifest as Namibia's National Biodiversity Programme (NNBP), for which an initial country study received funding from the United Nations Environment Programme (UNEP) (Barnard 1998). Both these programmes mark Namibia's status as a signatory of the 1992 United Nations Conventions on Desertification and Biological Diversity, thereby coordinating national environmental policy with international environmental priorities.

In pursuit of the ideal of 'sustainable development' (IUCN/UNEP/WWF 1980), and paralleling similar initiatives elsewhere in southern Africa, a third national programme of 'Community-Based Natural Resources Management', or CBNRM, ties environmental concerns to the country's rural development requirements. A key tenet of CBNRM is decentralisation, with 'communities' emphasised as the appropriate 'unit' of civil society to have decision-making powers over, and receive benefits from, wildlife resources. As elsewhere in southern Africa (for example, Zambia, Zimbabwe, Mozambique and Botswana), the Namibian CBNRM programme receives primary funding from the United States Agency for International Development (USAID), under the project title Living in a Finite Environment (LIFE). Additional donors are the World Wildlife Fund-United States (WWF-US), the United Kingdom's Department for International Development (DfID), and the World Wide Fund for Nature (WWF), as well as other development- and conservation-oriented institutions. Again, as elsewhere, implementation and facilitation is conducted largely by national non-governmental organisations (NGOs).

Given the exclusionary conservation policies of Namibia's colonial and apartheid past, a major achievement of the country's CBNRM programme is a policy designed to enable Namibian citizens to make some choices over, and receive income from, the management of wildlife. In communal areas this right is given to groups of people ('communities') who legally establish a 'conservancy' with a defined territory, a written constitution and a management plan.[2] A 'conservancy', in agreement with the Ministry of Environment and Tourism (MET), can then receive benefits from non-consumptive and limited consumptive uses of wildlife (such as tourism) (MET 1995a and 1995b). To date, some fourteen communal-area conservancies have been registered with at least thirty more developing (DEA Website

2001). The programme and policy have been recognised internationally as the most progressive initiative of its kind in southern Africa, with Namibia becoming the first country to be honoured for a *people*-centred environmental initiative with the WWF Gift to the Earth Award in September 1998 (Sutherland 1998). As such, the programme and its protagonists have attained something of an iconic status within conservation circles (cf. Brosius et al. 1998: 161).

In an attempt to generate debate regarding interrelationships between local, national and international interests and institutions, I present in this chapter an ethnographic analysis of a single incidence of resistance to the formation of a local conservancy. I emphasise that this relates to only one of many Namibian conservancies and is not a critique of the CBNRM programme as a whole, or of the national policy guiding conservancies. I will be focussing in part on documents generated by one NGO, referred to here as Integrating Conservation with Development in Namibia (ICDN), which is not its real name. My story relates to protest against ICDN, the NGO's response to it, and some of the dynamics underlying the rise of the dispute and the allegations surrounding it. These events occurred in an area of southern Kunene Region where I have carried out fieldwork on and off since 1992, my last visit being in 2000, shortly after the events related below.

The situation arose out of the formation of a specific conservancy, which wittingly or unwittingly, favoured one ethnic group, ovaHerero, over another, Damara. To clarify, ovaHerero (and Herero-speaking ovaHimba[3]) generally are cattle and small-stock pastoralists who speak a Bantu language (see, for example, Bollig 1998, Jacobsohn 1998). Herero are thought to have entered Namibia around the sixteenth century, when they crossed the Kunene River into the north-west of the country. A second large in-migration occurred in the eighteenth century, following which they are thought to have occupied the majority of the productive pastures of central Namibia (Werner 1998 and references therein). Werner (1998) argues that since the 'ethnicide' of the German–Herero war of 1904–1907, and within the constraints of the reserve and migrant labour system of the former South African administration, Herero have been consolidating their interests as cattle farmers by expanding their access to land.

Damara (and Nama), on the other hand, speak Khoekhoegowab, a 'click' language cluster associated with the Saan languages under the broader grouping of Khoesaan (Haacke et al. 1997). Damara are thought to have long associations with much of the central and western reaches of the territory now known as Namibia, and perhaps to be descendants of an extremely early migration into Namibia (Lau 1987: 4–5). Their categorisation as 'Damara' is, to some extent, a construct from early missionary and colonial ethnographic writings. These 'lumped' together a number of smaller exogamous groupings whose membership was reckoned with regard to interrelated concepts of geographic territory (*!hûs*) and lineage

(*!haoti*) (Lau 1987, Fuller 1993, Sullivan 2001). Historical records indicate that prior to processes effecting impoverishment among Damara 'groups' in the eighteenth and nineteenth centuries, those identified as such engaged in a wide variety of livelihood practices including trade in specialist items, livestock herding primarily of small stock, large-scale communal hunting, and plant gathering (Lau 1987, Fuller 1993).

The history of relationships between ovaHerero and Damara is complicated further by the region's pre-colonial and colonial history. Existing and dynamic interrelationships were impinged on by the influences of mercantile capital, the northward movements of Oorlam Afrikaner commandos from the Cape, and the arrival of European missionaries and colonists. Thus we can imagine a broad regional history of complex interrelationships characterised variously by cooperation, competition, conflict and distance, with shifts in the relative significance of these occurring through both time and space. My intention is to use the case material below to highlight the contemporary significance of a context of overlapping claims to land by people with different ethnic identities and historical experiences of inhabiting the area. The fact that a protest occurred with allegations made by, what some say is a vocal local faction concerned with a potential loss of power, and by what others might maintain is a disadvantaged group, speaks of political dilemmas in how the goals of community-based initiatives are implemented.

In analysing the case material I highlight two broad issues: first, some implications of the complex relationships which exist between a post-independence 'culture' of NGOs and the local, state and international locations in which they operate; second, the significance of ethnicity as a factor influencing access to negotiations and to employment positions created by environment and development initiatives. A closer consideration of ethnicity is important for understanding local conceptions of, and possible conflict over, rights to land and natural resources. Following Moore (1996), I feel it is important to highlight the way that historical experience, by the establishment of a particular conservancy influences who is elevated and who feels marginalised. Given the cultural dimensions of both of these issues, as well as the particular complexities generated by the locating of environment and development initiatives in diverse ethnic landscapes, I wish further to raise the possibility that ethnography might have a relevant role to play in making public a *range of views* on the unfolding of such initiatives.

A case story: Local attempts to bury an internationally-funded NGO

Figure 6 is a photograph of a grave. The inscription on the wooden cross at the head of the grave gives the name of an NGO ('ICDN') with the epitaph 'birth unknown, died 06.02.2000, Rest In Peace'. The grave is located directly outside the conservancy office of the settlement of Sesfontein/!Nani|aus. The office comprises a recently constructed wooden structure placed close to a large, spreading fig tree (*Ficus sycomorus L.*) growing at the largest of the settlement's natural springs or 'fountains'. This is a location laden with historical, economic and symbolic importance, as I outline below.

Figure 6   A symbolic grave constructed for the NGO ICDN on 6 February 2000. Photo: Sian Sullivan.

The grave was dug at the culmination of a 'protest march' which, together with a written petition, drew attention to a number of complaints regarding the establishment of a conservancy in Sesfontein/!Nani|aus and environs. The instigators of these events were the settlement's Senior Headman and the Sesfontein Constituency Regional Councillor (or Pastor). The petition formed the basis for a later document listing 145 names of Sesfontein inhabitants with fifty-five actual signatures, i.e., representing a substantial proportion of the settlement of Sesfontein's adult population of approximately 400 individuals (National Planning Commission 1991). The document also listed thirty-five names of inhabitants from Purros, accompanied by twenty-four actual signatures. Purros is a settlement located north-west of Sesfontein and has a separate gazetted conservancy. Significantly for some of the accusations and counter-accusations made in this case, it is inhabited predominantly by Herero-speaking Namibians.

This document was distributed to a number of international donors active in Namibia as well as to the Director of Resource Management of the Ministry of Environment and Tourism (MET). Specific accusations related to a number of employment and other practices by ICDN. These were said to favour external over local interests, to have exacerbated local conflict along ethnic lines, and to have played a role in the initiation and support of an emerging leadership challenge to the Senior Headman of Sesfontein. A particular accusation was that access to NGO and conservancy vehicles was influencing who was able to attend meetings and committee elections in favour of those supporting the NGO's views and practices. Needless to say, these allegations are controversial and are vehemently denied by the NGO. For the purposes of this paper, I wish to draw attention to two elements of this case. First, the fact that a small but significant group of protesters, anxious that their voice and interests were being undermined, decided to make public their concerns that an externally funded NGO was making impacts upon the texture of local ethnic and other relations. Second, that the NGO felt bound to respond to these expressions out of concern for the possible implications for the 'place' of NGOs *vis-à-vis* local, national and international environment and development interests.

The protest and petition were enacted some six years after the idea of communal-area conservancies was first debated in the area. The immediate catalyst was the removal of the locally elected treasurer of the Sesfontein conservancy committee following allegations of fraud. To illustrate the complexity infusing local interrelationships and circumstances, and drawing on personal correspondence and field notes, I include here some details and views surrounding this case. The conservancy treasurer was charged with being unable to account for some N$20,000 of donor money intended as conservancy staff salaries. He maintains that he paid the salaries, but that the receipts were stolen in an attempt to frame him. His lawyers have postponed the court hearing at least three times. Additional incidents make these accusations and counter-accusations more complex. For example, given a perception that ICDN considers that conservancies, as legally recognised local resource management institutions, should have the power to vet and approve PTO (Permission To Occupy land) applications, some think it significant that the accused treasurer already holds a PTO for a location within the proposed conservancy boundary. A PTO is necessary for the establishment of an economic enterprise such as a campsite or lodge for tourists; as such, they are an important resource given the focus on developing tourism as a means of increasing revenue from the non-consumptive use of wildlife. Further, the treasurer had been in frequent communication with a second conservation-oriented NGO operative in the area. This organisation was hoping to join forces with him in establishing a base-camp and tourism enterprise at the site of his PTO, from where they could increase the effectiveness of their monitoring of tourism and wildlife movements. Despite a shared interest in wildlife conservation and economic 'betterment' in the region, it is widely known that relations between the founders of these two NGOs have been poor in recent years.

After the removal of the treasurer from the conservancy office late in 1999, unease built within the settlement of Sesfontein. The office was closed. The NGO placed armed guards around the office from the currently Herero-dominated nearby settlements of Warmquelle and Otjimdagwe, as well as from within Sesfontein. This situation became extremely provocative for inhabitants of Sesfontein because the conservancy office is located in what is effectively the economic and symbolic heart of Sesfontein. The spring next to the guarded office feeds an old system of canals which irrigates gardens inherited nearly a century ago by Damara and Nama inhabitants of Sesfontein. The gardens and springs are considered a !anu or sacred place (Haacke and Eiseb 1999: 92–3), as a place of God (Elob). As such, the ensuing conflict located at this site has generated anxiety among those with a long history with Sesfontein and its gardens. Although the settlement's Senior Headman earlier agreed to this locating of the conservancy office, there is also some feeling that permission to construct the office was granted without consultation of those who have inherited land in the gardens.

In the ensuing weeks attempts were made to resolve the conflict situation. On 23rd February 2000, after the protest march, a meeting with ICDN was called by inhabitants of Sesfontein to iron out concerns prior to registering a conservancy in the area. Those present voiced the concern that the conservancy treasurer – 'the man who was chosen by the community' – either should be reinstated, or the conservancy office closed. Second, they demanded that those guarding the office should leave so that Sesfontein inhabitants could work their inherited gardens without fear. Third, attention was drawn again to perceived biases in negotiations concerning the conservancy. A particular complaint was that '[t]he weakening tool is the stream of vehicles which carry people to the meeting'. This is an accusation that donor-funded vehicles are made available to carry supporters of ICDN to meetings, while those with opposing views find it hard, logistically, to participate. The meeting was inconclusive, petering out after an official representing ICDN walked out. Some within the NGO took the view that the meeting was controlled by an aggressive and unrepresentative faction concerned to bolster their privileged access to resources. Whatever the 'truth' of the allegations and counter-allegations, it is clear that the introduction of donor monies to underdeveloped regions such as north-west

Namibia, even when manifest as rather modest capital resources (the location of a building, or the use of vehicles), can become the focus of intense political debate and attempts at appropriation. At a meeting the following day (24th February 2000), the Director of Resource Management (MET) recommended that a 'conflict resolution forum' be created in recognition of the social problems being caused by the process of conservancy creation.

Unsurprisingly, the petition and ensuing dispute have not been taken lightly by ICDN, the NGO facilitating conservancy formation. In March 2000 the NGO's directors circulated a four-page response to their donors and partner organisations. In this, the claims made in the petition are dismissed as 'unfounded and defamatory'. Explanations for the protest are attributed to an accusation that the NGO is 'being used opportunistically as a scapegoat in a power struggle between a small elite representing their own and/or their political party's interests at the cost of the rights of a majority of ordinary people'.

For our purposes it is interesting that the document simplifies many layers of complexity reflecting both the political ramifications of the NGO's activities and perceived allegiances within the region, and a context of identity politics based around ethnicity and assertions of difference. At one level, the statement by the NGO's directors appears to make the claim for an allegiance among the protesters to one of Namibia's political parties, the United Democratic Front (UDF). As successor to the Damara Council of the former 'homeland' of Damaraland's regional administration, and as the party of the Damara 'King' (Chief Justus ||Garoeb), this political party has strong Damara support. So, although the directors of the NGO do not refer directly to the ethnic dimensions of the dispute, the statement appears a response to accusations that the NGO has favoured Herero people in the area. The defence that the NGO offers here, that the accusations of the dissenting group can be explained by their links to an elite of a political party, is also interesting in that it deflects attention from the political stance of the organisation itself. As outlined above, the NGO and its employees, are also situated politically and clearly their work and actions are perceived locally as political.

In April 2000 ICDN began legal proceedings against those they considered the initiators of the protest and the petition. The charges, in a lawyer's letter from a Windhoek law firm, are that the petition 'contains allegations and remarks, which are not only utterly unfounded and false, but are vexatious and defamatory', causing '... damage to our client's reputation'. Without receipt of a written apology, the letter affirms 'instruction to institute an action ... in the High Court of Namibia for damages suffered' at the expense of the accused. The Senior Headman was threatened additionally with a charge of criminal assault. Given the circumstances surrounding the establishment of 'traditional' leaders under the country's pre-independence

South African administration (e.g., Gordon 1991), it might be appropriate to ask questions regarding the legitimacy or otherwise of their positions. I would suggest, however, that similar questions might be raised concerning the implications of an externally funded NGO invoking formal law against a traditional or customary leader, as well as against others voicing criticism and/or dissent.

In March and April 2000 a delegation of high-ranking leaders of the Damara Traditional Authority visited Sesfontein on a mission to diffuse the 'serious wave of discontent', thus highlighting the potential gravity of the local conflict situation arising from these events. Following a series of meetings, the delegation made a number of comments regarding the proposed conservancy and leadership dispute. They affirmed the legitimacy of the Senior Headman's leadership in Sesfontein, and recommended that allegations of fraud against the conservancy treasurer be dropped to help generate a climate conducive to cooperation. Regarding the conservancy, they suggested that it be dismantled in its current form. They also appeared to affirm the claim that 'most of the members were registered [for the conservancy] under the false pretext of getting game venison should they so register'.

In the period since these events the Regional Governor has been overseeing negotiations. A large document identifying all the objections made by the Senior Headman's supporters remains in circulation. In a follow-up exercise involving the drawing up of three 'counter-petitions', ICDN gained a mandate to continue working in the area from more than four hundred people from the broader territory affected by the proposed conservancy. Three international consultants contracted by DfID and WWF-UK completed an evaluation of ICDN's Kunene programme in November 2000. This approved the actions of the NGO and affirmed support from a number of 'communities', conservancies and Traditional Authorities. It is difficult to see, however, how this resolves the dispute emerging within the population affected by the particular conservancy which is the focus of this chapter. In December 2000 the conservancy office in Sesfontein/!Nani|aus remained closed and protected by an armed guard.

In summary, this 'story' highlights the intensely political and competitive atmosphere infusing conservancy formation, NGO and donor involvement, and CBNRM as a framework for environment and development initiatives. I acknowledge that it is not possible to represent all of the views of those involved in the multifaceted events and related documents detailed here, many simultaneous and possibly conflicting readings of which might also exist. In interpretation, however, and as outlined in the introduction, I will take up two issues: first, some of the political implications of interconnections between NGOs, new state policy, and a strengthening civil society which nevertheless may hold its traditional leadership in high regard; and second, the question of how identity politics based around ethnicity might have influenced the way that support was taken up and, in this case, contributed to emerging dispute.

NGOs: Measures of success, relationships to state policy and local allegiances

As in any profession, NGOs and donors are obviously under pressure to demonstrate successful performance, especially given the large amounts of financial resources involved.[4] Roe et al. (2000: 3) point out, however, that community benefits often are 'over-extolled by advocates' of 'community-based' resource management endeavours. Elsewhere I have suggested that the financial sustainability of CBNRM programmes in the absence of donor support is questionable. This is because it might be unlikely that revenue from wildlife and/or tourism will ever constitute a particularly large source of income for all members of a 'community' at household and individual levels, and that the costs of running conservancies as new wildlife management institutions are rarely factored into analyses of income (Sullivan in press).[5]

Here, however, I focus on the implications of the ways in which impressions of project and programme success frequently rely on a conceptual simplification of the socio-political contexts in which they take place. A primary component of this is a depoliticisation of the contexts and issues within and with which implementers, facilitators and donors are operating (for a case study *par excellence* see Ferguson 1990). Thus, power relationships between northern discourses of environment and development and local self-determination become lost. The tendency to embrace free-market principles as an accepted route to development and conservation (including tourism), for example, remains little problematised. This is so despite their highly political outcomes, including the commoditisation of both 'nature', and of the 'ecologically noble primitive' for consumers who tend to derive from the world's wealthy nations (e.g., Garland and Gordon 1999, Schroeder 1999: 367–70). At the same time, some of the real issues underlying poverty and a lack of options in communal areas tend to be avoided. In this case, history has dictated a context of inequality in land distribution, while current circumstances have stimulated in-migration of land-hungry pastoralists, often those who are relatively wealthy in terms of cattle herds. It might be expected that a continuing depoliticisation of issues relating to the global and national distribution of power and resources will contribute to future protest in development and conservation arenas. This is precisely what happened in the early 1990s when Kenyan Maasai pastoralists, in what has been described as a brilliant symbolic gesture that demonstrated their *power* over internationally valued resources, deliberately attacked elephant and rhino in Amboseli National Park in protest at the appropriation of their lands for conservation and commercial purposes (Bonner 1993).

NGOs have a particular place in these political dynamics. Elsewhere in Africa structural adjustment has necessitated a reduction of state expenditure on the management of public resources. In Namibia, attempts to raise user accountability for state-owned resources (e.g., Africare 1993), as well as a post-independence administrative weakness at regional and local levels, have paved the way for growth in the activity and significance of donor-funded NGOs. In some cases, donors and NGOs work in what seems like equal partnership with national government institutions in the formulation of new policy recommendations. For the environmental, and particularly wildlife, sectors in Namibia, the donor organisations of USAID, WWF-US and WWF, and particular individuals within these, certainly have been involved in close working relationships with the MET in the development of the communal-area conservancy policy (MET 1995c: 6; IRDNC n.d.). Although constructed as working for the common good, NGOs are not publicly elected or solely accountable to the recipients of their work. While providing expertise and funding and having an ability to redefine donor constraints and discourses, they also have to satisfy institutional and donor criteria for 'success'. This implies a degree of caution in such relationships, as recognised in Namibia in the MET's emphasis on drawing up memoranda of understanding with conservation-oriented NGOs (MET 1995c: 6).

In other words, current circumstances dictate that internationally funded NGOS have considerable power over national policy and programmes in situations where state resources are somewhat circumscribed. This is an important route by which the ideals of implementers and facilitators, as well as 'northern' business and consumer interests, can influence the conservation and development trajectories of 'the south'. The political ramifications of these processes tend to be written out in their recounting. While open resistance to such circumstances is rare for many reasons, NGO and donor inputs generally are moulded by local people into forms that inevitably depart somewhat from the implementers' original vision. Where the inputs of NGOs and donors demote local interests, or even *some* local interests, dispute and conflict may arise. In this case study, the perception of biased allegiances to particular local factions had an undeniable and significant effect in driving dispute. At the same time, the NGO, like other agencies, presents opportunities in the area which can be appropriated and manipulated locally, contributing further to a situation of fragmented support and resistance.

Ethnicity: Integral to, or excuse for, emerging dispute?

Murphree (1999: 4) asserts that '[e]xternal agents, inexorably, influence local governance by the power that they wield, the way that they intrude and the personal or functional alliances that they forge'. In the case-story, descriptions of alliances influenced by the presence of an NGO seem to be bound very much with ethnicity. Here, and in very general and essentialist terms, the donor-supported proposal for a conservancy is perceived as favouring one ethnic group, Herero, over another, Damara. In an official climate of nation building and the construction of a post-apartheid national Namibian identity, so-called political correctness might lead to a disinclination to discuss these complex but uncomfortable issues. Ignoring them or masking them, however, by referring to rural people as 'communities' or 'communal area dwellers' (i.e., without ethnic or other identity) does not change history or people's memories of experiences which may have marginalised them. Instead it might be that not enough sensitivity to, or transparency of, aspects of history, identity and culture vis à vis land have frustrated attempts to facilitate the registration of conservancies in the area, and have contributed to the resistance to NGO involvement described here.

In this area, invoking ethnicity in local arguments against ICDN may arise largely from a conceptual identification of the NGO's two directors with Herero-speaking people, given their personal histories of working in Kaokoland. That there is also some factual basis to the accusations, however, is suggested by a simple statistic: of 26 members of the conservancy committee at the time of protest, 18 were Herero while 8 were Damara. The proportions of inhabitants characterised by ethnic or language 'group' vary according to how boundaries for the conservancy are drawn and the extent of in-migration and other mobility. I would suggest, however, that this proportionally under represents Damara under any reckoning (e.g., National Planning Commission 1991).

It is important to emphasise, however, that the situation is not one of a simple split between two ethnic groups, exacerbated by the interactions of an outside NGO. In fact, different Herero groupings were very much involved with and in support of the two Damara factions within a Damara leadership dispute. This reflects the shared and intertwined histories of Damara and Herero in the area. Thus the Kasaona Herero 'group', associated with Warmquelle, Otjimdagwe and Ganamub settlements, aligned themselves with the Damara faction disputing the leadership of Senior Headman Gaobaeb. The Kasaonas in particular were seen to have benefited disproportionately from organisational efforts surrounding the proposed conservancy. Senior Headman Gaobaeb, on the other hand, was supported by a second Herero group associated with the Kangombe family, also from Warmquelle settlement. These two Herero groups have been in conflict for some time, and current disagreements provide yet another forum for playing out their own dispute.

These cross-ethnic interconnections call into question why it is that the dispute has been framed locally as a rather simple case of the NGO working with one ethnic group over another. I think that the significance of ethnicity in the case story to some extent becomes clearer only once the broader context is brought into focus. Damara and Herero, as well as other 'groups' present in the area and the wider north-west region as a whole, have a long history of 'muddling through' in their social and economic interactions. While ethnic and cultural 'difference' has been part and parcel of these interactions, they have not prevented mutual respect, intermarriage, and shared regional trade networks, involving extensive travel, for items produced by each other. I have often visited the cluster of huts in Sesfontein inhabited by Suro, my field assistant, and her family to find several traditionally clad ovaHimba visiting and socialising – perhaps exchanging perfume plants obtained from near the Kunene River in the north (*!garib sâi*) for the valued local tobacco grown by Damara in the gardens at Sesfontein. These multifarious interactions and inter-relationships have been forged under political and economic circumstances which, as a consequence of German colonial rule and the ensuing South African administration, were not of their own choosing.

Official changes in access to land, combined with input by NGOs, however, seem to have set in motion a dynamic that has tended to impinge on Damara rights and access to decision-making forums. A history of such dynamics perhaps shapes Damara sensitivities to recent and current changes, particularly a sense that in-migration may amount to loss of their land and historical places. This is a process that has been underway since at least the 1970s with the redrawing of administrative boundaries and the creation of 'homelands' following the recommendations of the infamous Odendaal Report (Government of South Africa 1964). Under this, Warmquelle (|Aexa|aus), which had been inhabited and worked by Damara people from at least before German colonial rule, became part of Opuwo District to the north and as such was re-created as part of a Herero/Himba constituency, i.e., as located in the Kaokoland 'homeland' for ovaHimba. Damara people inhabiting Warmquelle/|Aexa|aus were moved slightly south to Kowareb in what was designated 'Damaraland' (Sullivan and Ganuses, in prep.). Reportedly, it is only since this time that the Kasaona and Kangombe families, now so important in the local politics of the area, settled in Warmquelle.

Fears that such processes of settlement and land loss will be repeated perhaps underscore the Damara opposition to the Sesfontein conservancy in its current form (although elsewhere Damara or others have invoked membership of a conservancy in attempts to keep out herders in search of grazing [Inambao 1998, Shivute 1998]). Uncertainty regarding claims to land is compounded by Namibia's post-independence constitution which provides for all Namibians to move to wherever they wish on communal land with the proviso that they 'take account of the rights and customs of the local communities living there' (Government of the Republic of Namibia 1991: 28–29). Without an institutional basis for monitoring the effects of such movements or for protecting the rights of existing residents, however, this otherwise liberal context can be disempowering for some. This is especially so given a situation where options for movement tend to be greatest amongst the wealthy (Rohde 1993, Sullivan 1996), and where ethnicity as a major axis of difference tends to conspire against certain groups (Botelle and Rohde 1995, Taylor 1999, Twyman, in press).

This is precisely what appears to be happening in the Sesfontein area where relatively wealthy (in terms of livestock) Herero pastoralists seem to be increasing in number. For example, Kowareb in the 1991 census registered no Herero households (National Planning Commission 1991) and its largest herd of cattle in 1992 was of 25 (personal field data). Now there are several Herero homesteads within the settlement and large Herero cattle herds roam the village and its surrounding environment. In-migration, of course, can bring benefits including adding vibrancy to local economies and increasing employment opportunities. But it can also instil a sense of loss of control and displace people whose previous claims to, and knowledge of, the landscape are overshadowed (as noted for the 'Saan' elsewhere in Namibia, e.g., Botelle and Rohde 1995, Thoma and Piek 1997). As Andreas !Kharuxab, longstanding Damara headman of Kowareb, describes:

> There are too many livestock of people that we do not know in this area. ... People of other areas move in with their livestock: people of Opuwo District, people of Hoaruseb area. The people move in here with their livestock and their animals are many.

> When this land became independent the government said that everyone can move where they want to. ... But those words bring conflict. Look, if a person moves because of drought and asks if we can please help him because his livestock are dying, and he comes in to another person's area, then we give him space [for herding]. But if at the end of the year when the rain falls at his place he doesn't go back there ... then his livestock are here all the time. The government should look at these things. If the person moves here because of drought, and if the rain falls at his place at the end of year, then he should take his livestock back there and farm at his place.

> Recorded interview, 12th May 1999

These dynamics, together with the specific conservation priorities of CBNRM initiatives, act further to overlook, or even displace, the denseness of Damara memory and knowledge of landscape, of specific resources, of ancestral land-areas or !hûs, and of place-names (Sullivan 1999, 2001, in prep.). As Moore argues, 'Historical patterns of access to resources and exclusion from them mold cultural understandings of rights, property relations and entitlements: in turn, these competing meanings influence people's land and resource use' (1996: 128). This brief background to the current dispute suggests that this is indeed the case here. Given historical circumstances, poor representation on the conservancy committee combined with observations of perceived bias in the behaviour of one NGO perhaps contributed here as a proverbial 'last straw' in provoking dissent.

Conclusion

This book emerged out of a concern with 'ethnographies of environmental underprivilege'. Here I have related an incident of local protest to dominant and donor-funded NGO activities in the conservation and development arenas. I have suggested that it can be understood as: 1) an expression of resistance in part to what remains a multilayered situation of national and global inequality in the distribution of power and resources, and 2) a protest against longstanding circumstances which are considered to undermine the leadership and other rights of a particular 'group' in the area, and no doubt prompted by their fears of a loss of whatever powers they currently have.

In presenting this case, however, I am driven to consider a further issue about the role and responsibility of ethnographers as chroniclers of interactions between local people and international environment and development institutions. Lurking behind this ethnographic account, as readers may have sensed, is the threat of litigation against words and interpretations that are not welcome in the business of international conservation. This draws us into an ethical issue: should ethnographers present and publish minority points of view held by those who fit awkwardly into the regionalist plans of organisations?

Academics can appear rather sanctimonious in their writings regarding environment and development: preaching from the margins while avoiding 'getting their hands dirty' with the 'real' work of implementation. In consequence, the possibility of engaging those who 'matter' in debate – policy-makers, donors, implementers, facilitators and local people – is compromised, especially if those 'on the ground' feel under attack by academic work. In response to my own writings on this region I have experienced resistance and abuse, and threats of libel suits. Some of the responses may sound familiar, or perhaps even have salutary value, to other academics exploring similar issues. For example, it has been said that I can 'afford to be so fiercely critical of others ... [because] in academe the stakes are so low', that my 'arrogance leads [me] to raise serious questions about the integrity and honesty of people working in the field', and that, like 'the news media and the harlot' I exploit my situation in the academy as one of 'power without responsibility'. Several respondents have taken offence at observations I have made of possible gender implications embodied by current emphases of CBNRM activities – stating, for example, that I employ 'smear tactics' in order to 'enhance [my] own reputation as a slayer of all that is white, male and involved in conservation'. Perhaps, needless to say, these ignored my reference to concerns in the wider literature – expressed by men as well as women – regarding constructions of masculinity in a southern African context (e.g., Mackenzie 1987, Ellis 1994, Carruthers 1995). To provide a rather flippant example that these remain real issues, I was told on my last visit to Namibia that an employee of the Directorate of Environmental Affairs (MET), on reading an earlier paper (Sullivan, in press), responded with the comment 'That girl writes as if she doesn't have a boyfriend'! Finally, in some contexts my work has been attacked by my association with people attached to one side of a local leadership struggle. I am considered as being 'naïvely manipulated by a local political elite' (personal correspondence) due to my field assistant's support for Sesfontein's Senior Headman.[6]

Others, however, have welcomed alternative views. A representative of RISE (Rural People's Institute for Social Empowerment), an indigenous NGO involved with the Namibian CBNRM programme, commented, for example, that she was 'very animated' by the paper. Moreover, she stated that she 'felt that the LIFE staff should know that they are working with local partners who are thinking critically and reading material beyond what they as LIFE provide', in which the programme tends to be portrayed 'as an enormous success'. Ironically, in her enthusiasm she emailed copies of my paper to various people in local NGOs as well as to the Director of the LIFE programme, not realising that the paper had already elicited an attack from these quarters.

In other words, the uptake of ethnographic and other analyses is complex. Research is either declaimed or utilised depending on an individual's or organisation's needs and positions, never mind the 'truth' of any claims made in such writings. So, two factors might compromise the ability or otherwise of academic analyses to perform some sort of 'public service' (Gordon 2000): first, what might be considered an in-built resistance to external criticism because of institutional needs to demonstrate success; and second, because critique tends to be framed in the rather confrontational manner typical of much deconstructive academic writing, thereby pushing implementers into defensive attack. Without question, implementers uncomfortable with independent analyses have the right to respond to critique. In academia there are accepted channels for doing so, and review and critique are expected as part of debate. However, and although it might be appropriate to review the ways that 'we' write, it should be emphasised that academic researchers and others also have the right to voice critique and alternative views. Perhaps this becomes particularly important when large organisations have the resources to silence weaker minority groups (and academic researchers) through threats of litigation.

Beyond these ethical issues surrounding rights to public representation, there is yet another dimension to this story which raises the stakes surrounding concepts such as 'freedom of speech', advocacy and duty, and their employment in the complex interrelationships guiding a globalising of conservation effort. This is that African land users potentially hold substantial power over species which, in today's world of a commoditised nature, are highly valued in international circles. Elsewhere, when the state wishes to manipulate land-use but where land holders constitute a powerful lobby, realistic alternatives are offered in return for the opportunity costs of sacrificing productive land-uses aspired to by the land holder. Given the large amounts of monetary resources flowing from the wealthy countries of the north to wildlife-rich areas of the south, perhaps it is time that some form of direct payment to rural Africans is considered as a means of offsetting the opportunity and other costs that they pay for retaining dangerous animal wildlife and conserving valued habitats on 'their' land. Such suggestions have been made by economists Simpson and Sedjo, who argue that instead of investing in activities felt to be related to conservation but which require external funding, 'donors might attempt to pay for conservation performance directly' (1996: 242, cf. Norton-Griffiths and Southby 1995, Norton-Griffiths 1996). Although they acknowledge that instituting the mechanisms whereby this might occur would be an administrative nightmare, they point out that establishing commercial ventures where conservation goals are achieved indirectly are little better. Were tenure over land to be clarified for Namibia's communal-area residents, such an approach would not necessarily detract from a 'community's' potential to negotiate lucrative partnership agreements with foreign investors, or even from using the current conservancy framework and policy to do so. The allocation of donor monies directly to the individuals who inhabit valued environments would perhaps be more empowering, would certainly bring greater per capita returns and thereby perhaps make possible a greater range of livelihood and lifestyle options, and, if successfully linked to conservation activities, could bring the conservation returns that are hoped for.

Acknowledgements

I owe a large debt to David Anderson and Eeva Berglund for their advice and input, and to their and the publisher's patience while I was revising the chapter. As always, Suro Ganuses provided invaluable assistance and good company while in Namibia. I have drawn gratefully on discussions with Dan Brockington, David Simpson, Robert Gordon, Kathy Homewood, James Fairhead and Gerard Sullivan in the writing of the paper. I also acknowledge, with appreciation, detailed comments on an earlier draft from Margie Jacobsohn, Blythe Loutit and Eugéne Marais. I hasten to add that the views contained herein are my own. The paper is written as part of a British Academy Post-Doctoral Fellowship, with recent fieldwork funded by a Small Grant from the Nuffield Foundation in 1999 and a Small Research Grant from the British Academy in 2000. I gratefully acknowledge the support of these institutions.

Notes

1   GTZ = Deutsche Gesellschaft für Technische Zusammenarbeit of the Federal Republic of Germany.
2   Communal areas refer to land owned by the state but designated to be used and managed by African farmers on a communal basis, in contrast to land owned under freehold title by settler farmers (primarily Afrikaans-speaking). While the latter are in a minority, they hold the majority of the most productive land in southern and central Namibia.
3   A term which distinguishes Herero who remained in Kaokoland, and who later fled to southern Angola to escape the violent livestock raids of Oorlam Afrikaner commandos from the Cape, from those who migrated southwards.
4   The major USAID-funded national CBNRM programme (LIFE) received US$25 million from 1993 to 2000 (Callihan 1999: 6–7). Of this, US$14 million was channelled to 'ICDN' between 1992 and 1999 for work in regions surrounding Kunene (Durbin et al. 1997: 28). A further US$12 million from USAID was approved to carry the Namibian CBNRM programme from late 1999 to 2004 (Callihan 1999: 6–7). ICDN also received Swiss Francs 2,794,550 from WWF-International towards its work in Kunene Region between 1996 and 2001 (Jones 1999: 76).
5   Some of Namibia's 'flagship' conservancies have generated substantial amounts of income. As has been pointed out to me, for example, 'Torra conservancy has weaned itself off donor support and manages its conservancy, office, vehicle and staff of seven from its own income, plus has made a profit for its membership, with more than N$300 000 in the bank' (personal correspondence). This is rather exceptional, however, when considered in relation to the country as a whole. Moreover, these figures do not factor in the capital that has been required to establish the conservancy over the last six years or so, via both NGOs and investment by private enterprise.
6   My field assistant is related by kin to Sesfontein's Senior Headman. It might be added, however, that she is more closely related in terms of kinship to his second-in-command, who has challenged his position, a dispute that is considered to be variously supported by the NGO, not least due to participation in the dispute by one of its senior employees.

5

ENVIRONMENTALISM IN THE SYRIAN *BADIA*
The Assumptions of Degradation, Protection and Bedouin Misuse

Dawn Chatty

The Syrian *Badia*,[1] the vast semi-arid steppe land which makes up nearly 80 percent of the state's land mass, has been at the heart of centuries of significant political struggle between pastoral Bedouin tribal authority and the power of the centralised state. In the early twentieth century, with the end of the Ottoman Empire and the imposition of a League of Nations Mandate, the French authorities first set about encouraging the Bedouin to govern themselves. But after finding that inter-tribal raiding and skirmishing were affecting France's development plans, they vigorously pacified the area. For the last half of the twentieth century, the independent state has sought to complete the dismantling of the Bedouin tribes of Syria. Failing to successfully coerce Bedouin to settle, the government undertook to strip its leadership of all power and authority. The *Badia* was nationalised and all tribal holdings ceased to be recognised in 1958. By the 1960s, the language of environmental degradation, desertification and overgrazing entered the political vocabulary of technicians, diplomats and politicians alike. Most technology transfer to Syria was aimed at taking over greater areas of the *Badia* and converting them into important agricultural crop-producing regions. Today, the Bedouin, having been pushed back ever deeper into the *Badia*, are regarded by the state as destroyers of their own homelands, accused of overgrazing precious shrubs and grasses, hunting the gazelle and oryx into extinction, and carrying no concern for or knowledge of the sustainability of this fragile land. However what emerges in this study is that a political understanding between government and tribal leadership has supported the continued existence of alternative systems of land use among the Bedouin which is unofficially tolerated, but officially denied.

Historical background

Throughout the eighteenth, nineteenth and early twentieth centuries the Syrian *Badia* was the source of significant environmental contestation with use rights for graze, browse and water passing from the hands of the weaker pastoral nomadic tribal groups to those Bedouin tribes with military superiority. The authority of the Ottoman Empire hardly reached this region and its border area with agricultural lands, the *Ma'moura*, was the site of constant skirmishes between central authority and pastoral nomadic tribes (see Figure 7).[2] When the central authority of the Ottoman State was strong the Bedouin were generally pushed back from the borders of the *Ma'moura* deep into the *Badia*. Weakness or distraction of central authority would generally mean that the Bedouin could expand into the *Ma'moura* and sometimes beyond into well-established agricultural zones. With the end of the Ottoman Empire and the imposition of a League of Nations Mandate, the French authorities set about encouraging the Bedouin to govern themselves, perhaps influenced by some romantic eighteenth and nineteenth-century image of the 'noble savage'. Bedouin tribal leaders were supported by a special French administrative unit, the Contrôle Bedouin, which was outside the jurisdiction of the French civil administration. This unit encouraged traditional Bedouin law and conflict resolution to operate in the *Badia*. Occasionally, skirmishes overspilled into agricultural areas governed

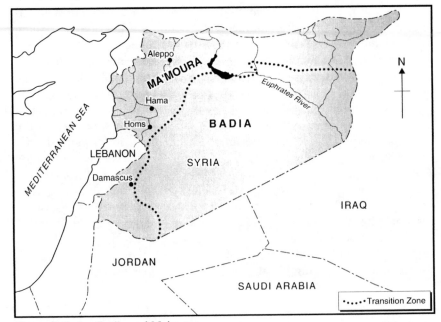

Figure 7 Syrian Badia and Ma'moura

by a separate colonial administration. But as long as French interests were not affected, the Bedouin were informally allowed to operate as a de facto 'state within a state'.

However, with the discovery of oil in the region, the French Mandate power became concerned with protecting a potentially important international investment. After finding that inter-tribal Bedouin raiding and skirmishing were affecting the laying and protection of oil pipelines from the interior to the Mediterranean coastline, the French reversed their original policy and vigorously pacified the area, stripping the tribes of their semi-autonomous status, and coopted the leadership into the urban elite of Damascus, Hama and Aleppo. This was accomplished largely through grants of private ownership of vast swathes of the common tribal grazing areas of the *Badia*, voting rights in Parliament, privileged access to foreign education for the sons of Bedouin leaders, and significant monetary compensation (France 1923–1938).

The establishment of the independent nation-state in the late 1940s and 1950s saw the continuation of several decades of sustained effort to control and break down pastoral tribal organisation. Much of the tribal leadership was coopted into the elite urban political scene. Land holdings, once held in common, were increasingly registered in the names of tribal leaders and converted into farms. The Bedouin tribes of Syria, and Northern Arabia in general, struggled with two opposing forces: one compelling them to settle on the edges of the desert and engage in marginal agricultural production, the other forcing them to move away to seek multi-resource livelihoods and pastoral subsistence across several national borders (Abu Jaber et al. 1978, Lancaster 1981, Chatty 1986, 1990). In September 1956, after several years of continuous skirmishing in Homs, Hama and Aleppo, the government summoned all the major tribal leaders to Damascus. This was ostensibly an effort to arbitrate the conflict between the tribes and sign a 'peace' treaty. The occasion was also used as the first official and formally documented step in dismantling any government recognition of a population which had no fixed abode, did not receive any state services, and was not accessible to control either by police forces or security services. Failing in its efforts to entice Bedouin to move out of the control and orbit of their leaders, and to settle on farms in the *Ma'moura*, the government undertook to strip the Bedouin leadership of all power and authority. In 1958 the *Badia* was nationalised and all tribal holdings ceased to be recognised by the state, the entire area coming under 'state ownership' (Masri 1991, Rae 1999). With this measure the government believed it had completed the dismantling of the Bedouin tribes which had begun nearly fifty years earlier with the French neo-colonial administration.

Formal state transformation of Bedouin land use

The 1960s were a period of strenuous government land reform, including not only the formal seizure of all commonly held tribal land but also the confiscation of the large tracts so recently awarded to individual Bedouin tribal leaders as private holdings. Much of these confiscated holdings were given to urban merchants, favoured politicians, and entrepreneurs for large-scale industrial development of cotton and wheat production in the *Ma'moura* and other less arid areas of the *Badia*. Following a three-year-long drought in the early 1960s, in which over two million sheep died, the government instituted a programme to alleviate the problems caused by this ecological disaster. The government set about reviving the livestock industry without also restoring authority to tribal leaders, or tribes to their traditional lands. Terms such as environmental degradation, desertification and overgrazing came to be used by technicians, diplomats and politicians alike when discussing the Bedouin and their use of the *Badia*. Development aid and technology transfer to Syria was aimed at taking over greater areas of the *Badia* and converting them into important agricultural crop-producing regions. A United Nations-sponsored project was set up to revitalise the pastoral sector of the Syrian economy, but not the structure of its society. Its foremost goal was to stabilise the mainly pastoral livestock population. This proved extremely difficult since the agricultural and livestock technicians running the project – mainly trained in the West – did not understand Bedouin methods of animal husbandry.[3] In turn, the Bedouin had no trust in government – especially in light of the recent confiscation of grazing land, and the explosive expansion of agricultural development over nearly a third of the best rangelands of the *Badia* (Al-Sammane 1981: 32).[4]

International experts assigned to the Syrian government declared the *Badia* degraded due to overstocking and poor indigenous range management practices (see ILO 1964, FAO 1965). At the same time, government and international agencies concerned themselves with the revival of the sheep livestock industry. In Syria and much of the developing world, development and conservation efforts have been largely based on the assumption that human actions negatively affect the physical environment. Problems such as soil erosion, degradation of rangelands, desertification, and the destruction of wildlife have been viewed as principally due to local, indigenous misuse of resources. Recent studies have clearly shown that models of intervention developed in the West have been transferred to the developing world with no regard for the specific contexts of the actual receiving environments or peoples (e.g., Sanford 1983, Anderson and Grove 1987, Manning 1989, Benkhe et al. 1991). The Western, urban notion of wilderness as untouched or untamed land, for example, has pervaded conservation thinking and been broadly exported to the developing world. Parks and nature reserves in many parts of the world were created by first evicting indigenous people. What is now beginning to be recognised, however, is that the very ecosystems which conservationists wish to protect from people were, in part, maintained if not created by the indigenous human inhabitants and their livestock.[5]

Without any empirical studies or baseline data from which to judge, first one and then other international development agencies joined the government in declaring the *Badia* severely degraded. This in turn led to special programming and project development based on Western philosophies and technologies derived mainly from the United States and Australia. Foremost among these were the concept of sustainable yield and the goal of improved productivity. These concepts originated in North America and were rapidly adopted in Australia. They were derived from the experience of cattle ranching in the first half of the twentieth century, where both the cattle and the land upon which they grazed were part of a system of private ownership. Furthermore, the land had formal, inflexible borders which could not be contested or altered in anyway. Since that time – nearly fifty years on – policy-makers have defined the major concern of pastoral regions of the developing world to be similar to that of ranchers in the United States and Australia, namely overstocking leading to certain ecological disaster. In this view the problem has a technical solution – destocking. However the central assumption underpinning these sets of assumptions is that pastoral ecosystems are potentially stable and balanced, and become destabilised by overstocking and overgrazing. This bias has led to the establishment of a multitude of development projects that promoted group ranching, grazing blocks and livestock associations. These schemes have failed, however, leading to a fundamental questioning of the basic assumptions underlying this tradition of range management. Behnke et al. (1993) have admirably shown that pastoral systems are not equilibrium systems. Instead they are continuously adapting to changeable conditions, and their very survival depends upon this capacity to adapt. It is, in fact, the 'conventional development practices themselves that are the destabilising influences on pastoral systems, as they have prevented traditional adaptive systems from being used' (Pimbert and Pretty 1995: 5).

In Syria, the first large-scale international development project in the 1960s focussing on livestock and range 'rehabilitation' ran into trouble within a few years. After four years of poor results, a handful of specialists with the Food and Agriculture Organisation (FAO) and the World Food Programme (WFP) launched a campaign to convince the concerned agencies of the importance of studying the human factor. They argued that unless development programmes were in harmony with the customs and ways of life of the pastoral populations, the rangeland development scheme would fail. Bedouin as well as government cooperation was required in order to solve the problem which the government perceived was simply one of rising livestock numbers.

In 1967, Draz (1977) came to appreciate that the traditional Bedouin system – which operated informally – was alive and healthy in spite of government efforts to impose a modern, Western system of management. Given the poor government project results, he recommended that the Syrian government reconsider its position vis-à-vis the human population and revive the Bedouin tradition of *hema*[6] and thus return control over range conservation and management of grazing lands back to the Bedouin. This recommendation was as a response to what was perceived by government to be a 'tragedy of the commons', an open-access free-for-all. The government and its international advisors had assumed, as was common throughout international development circles, that with access to the grazing lands of the *Badia* no longer controlled by the Bedouin, as a result of the nationalisation of all tribal land holdings a decade before, these areas were naturally becoming degraded from overstocking and overuse.

The assumption in international circles and in government was, of course, that the 'nationalisation' programme had actually taken effect and access to pasture in the *Badia* was actually open and free to use on a first-come, first-serve basis. In fact, the Bedouin continued to use the *Badia* as they had done for centuries before, negotiating between and within tribes for access to resources. The basis for this system of land use had been undermined by the recent government decrees, but it had not been destroyed. Draz's recommendations for a return to a system of communal ownership was an indirect recognition of the de facto existence, if technically illegal, of an alternative tribal system of resource allocation. His suggestion appealed to the Syrian government's socialist orientation and the proposal was accepted.

After several years of trial and error a programme of *hema* cooperatives was implemented in the early 1970s whereby block applications by tribal units for control over their former traditional grazing lands were generally granted by the government. Power and responsibility within a cooperative was assumed then to be in the hands of cooperative members, in some cases mainly made up of one tribe. Its members were assumed to have a participatory role in the programme, taking part in meetings to determine the price of animal fodder, feed supplements, and in its earlier days, credit facilities for members.[7] Some tribal groups accepted this government administrative superstructure and restricted access to tribal lands. Others, however, did not and many moved away to Saudi Arabia and Jordan. By the mid-1990s the government claimed to have over 400 *hema* cooperatives covering approximately 5 million hectares of the *Badia* (Syrian Arab Republic 1996).

Throughout the 1960s and 1970s, the government was concerned mainly with raising livestock numbers (FAO 1972a). Each failure to meet a production goal was blamed on the degeneration of the grazing land, the loss of grass and ground cover as a result of Bedouin overstocking and overgrazing of the *Badia*. The government response to these failures was to create numerous research stations, fencing off more grazing land and restricting access by Bedouin to greater swathes of the *Badia*. The first of these was at Wadi Azeeb between Homs and Hama. This research station was set up on the land confiscated from Malawi and Hadiidiin tribes during the attempt to find a peaceful solution to their long-standing feud in 1956 (Rae et al. 1999). Besides introducing exotic sheep to improve the already superior local breed, the fat-tailed Awassi, the station management fenced off, seeded and planted shrubs common to Australia. The most prominent of these were the *atriplex* species, a particularly drought- and saline-resistant plant, but which local livestock found uninteresting. No scientific study was ever undertaken to determine whether or not the rangeland was overgrazed or suffered from 'desertification'. Only in the late 1990s have studies comparing aerial photographs of the *Badia* in the 1930s and the 1990s been undertaken, with surprising conclusions.[8] The government, advised by international experts (FAO 1972b, Peterson and Van de Veen quoted in Rae 1999: 8), continued to set up research stations, fencing off tens of thousands of hectares, in its efforts to protect and conserve flora which they assumed was under threat from local users.

Ironically, these projects also attempted to integrate the Bedouin membership of the *hema* cooperatives into the government's conservation projects. The rationale behind these measures and pilot projects was to attempt to rehabilitate rangelands, protect threatened plant and shrub species and stop the incursion of thorny bush. The government hope had always been that the Bedouin would appreciate the benefit of fencing and exclusion and be inspired to do the same on traditional land holdings under cooperative 'management'. Unfortunately this did not happen. Instead, the Bedouin expressed resentment at traditional common lands being confiscated for government experiments, which they perceived brought them no tangible benefits (Chatty 1995, Roeder 1996).

Despite numerous ups and downs caused by changing legislation, and inadequate restraint on the spread of agriculture into the *Badia*, the current situation, which allows Bedouin occasional voice in the government organisation of the *hema* cooperatives, is an improvement over the rigid government regulatory schemes of the 1960s. The two factors underpinning the formal organisation of the *hema* cooperatives were flexibility and a complicity in accepting the continued operation of traditional Bedouin systems of exploitation and marketing. These fundamental factors have resulted in a national programmes which *de jure* strips Bedouin systems of land use and resource allocation of all authority, but de facto recognises that these traditional operating systems, flawed as they might be, do exist.

Western conservation philosophy in practice in Syria

In the past decade, government concern has shifted to global issues of conservation, land mismanagement by indigenous populations, and the extinction of wildlife through the uncontrolled hunting of local residents. Much like East Africa, Syria is now playing the conservation card in a bid to enter the modern fraternity of nations concerned with biodiversity. Again like East Africa and other parts of the world where the philosophy of conservation protectionism has been exported, the Syrian government blames its indigenous Bedouin populations for the sins of its elite classes and other powerful ruling groups.

Government-organised parks and protected areas first made their appearance in America and Europe during the last century. Significant areas of land were set aside as wilderness, to be preserved 'untouched by humans', for the good of humanity. In 1872 a tract of hot springs and geysers in northwestern Wyoming was set aside to establish Yellowstone National Park. The inhabitants of the area, mainly Bannock, Crow, Sheepeater and Shoshone native American Indians, were driven out by the army, which took over management of the area (Morrison 1993).

In the United Kingdom, conservationists, mainly foresters, stressed that the public good was best served through the protection of forests and water resources, even if this meant the displacement of local communities (McCracken 1987:190). This expertise and philosophy was exported abroad to all of Great Britain's colonial holdings. Now, a century later, most national parks in Latin America, Asia, Africa and the rest of the developing world have been and, to an extent still continue to be, created on the model pioneered at Yellowstone and built upon by the early British colonial conservationists. The fundamental principal of operation remains to protect the park or reserve from the damage which the indigenous local communities inflict.

As was the case in the formation of Yellowstone National Park, armies or colonial police forces in Latin America, Africa, Asia and much of the developing world have been employed to expropriate and exclude local communities from areas designated as 'protected', often at great social and ecological costs. Forced removal and compulsory resettlement, often to environments totally inadequate for sustainable livelihood, were common practices.

Accompanying this forced removal was the view that indigenous people who rely on wild resources are 'backward' and so need help to be developed.[9] Occasionally, the land upon which these indigenous people lived, was deemed better used for modern agricultural practices. The situation of the Maasai in Kenya and Tanzania is another example (Jacobs 1975, Lindsay 1987). In 1904, in an effort to pacify the Maasai and to clear preferred land for European settlers, the British government created the Northern and Southern Maasai Reserves. Subsequently, over the next ten years, the Colonial government abolished the Northern Reserve and forced its resident population to move, effectively denying them access to much productive rangeland. It prohibited all hunting of wild animals on the reserve. These reserves served the purpose of preserving primitive Africa where 'native and game alike have wandered happily and freely since the Flood' (Cranworth 1912: 310 quoted in Lindsay 1987: 152).

In 1992, Syria negotiated funding for a project to rehabilitate rangeland and to establish a wildlife reserve in the Palmyra *Badia*. A choice area, one of the few remaining unrestricted camel grazing terrains in Syria, was selected by an international conservation expert as the ideal site for the reintroduction of the Arabian oryx and gazelle. A two-metre-deep trench was dug out by bulldozer to define a rectangular area 75 miles long and 25 miles wide. It was called *Taliila*, and its scar is visible on satellite images. This trench effectively prevented access to *Taliila* by Bedouin with their herds and trucks. The Syrian request for international funding was accepted and the Food and Agriculture Organisation (FAO) was drawn into the operation of the project as it appeared to have a development focus of improving food security. The project proposed to address three interrelated issues: diminishing grazing land, disappearing wildlife, and increasing requirements for supplemental feeding of domestic herds. As before, the assumptions underlying these objects were that the Bedouin had overgrazed the *Badia* thus diminishing grazing land, and had over-hunted large mammal species like oryx and gazelle, thus contributing to their extinction in the region. Furthermore, they were overstocking their herds, thus requiring feed supplements.

The project also proposed to restrict the Bedouin from land over which they had both formal and informal usufruct. At the completion of the third year of the programme, it was intended to dispossess the Bedouin altogether from the area earmarked for the animal reintroduction effort. The project proposed to incorporate some of the land holdings of three *hema* cooperatives into protected ranges, to set up restrictions on access by Bedouin and their domestic herds, and to run a programme to introduce new plant species. After two years of this three-year effort, the project expected to have obtained a high enough ' forage production ... to enable domesticated animals and wildlife to live in harmony on the land' (FAO 1995: 7). In the third year of this project, physical boundaries were to be established and 'the reserve will only be devoted to wildlife grazing' (FAO 1995: 7). In other words, at the close of the project, the Bedouin and their herds were to be completely excluded from an important area of rehabilitated rangeland. At no time were the Bedouin consulted or informed about this looming dispossession.

The programme is now in its second, three-year cycle and, not surprisingly, many of its goals have not been achieved. Although there is a recognition that the 'integration and effective collaboration of the beneficiaries to the programme' is required for sustainability, no visible effort has been made in the technical description of the project to incorporate the Bedouin in its planning, development or implementation. Instead, or in addition, the project document specifies that similar schemes in Saudi Arabia and Jordan will be studied in order to increase the likelihood of success in Syria.[10] The indigenous Bedouin population, however, are only to be involved peripherally in the analysis of field data. Representatives from the *hema* cooperatives are to be involved in the data-recording process and in the discussion of results in order to 'develop their awareness on environmental protection' (FAO 1995: 11). The wildlife reserve, *Taliila*, which received eight oryx from Jordan's Shawmary Reserve and sixteen gazelle from Saudi Arabia, will remain the home for these animals for the foreseeable future. The Bedouin have been excluded from any role in the planning and management of the reserve, and even the four 'local' guards at the entrances of the reserve are drawn from the town of Palmyra.[11]

What is striking from this inventory of 'facts' is the short memory of government. The lessons learned in the 1960s appear to have simply been forgotten: pastoralists cannot be separated from their animals or from their common grazing land. Furthermore, the underlying assumption of this project seems to be again turning back to the old bias that it is pastoralists that are overgrazing or overstocking, and that the solution is to reduce herd numbers and restrict their access to land in order to protect its carrying capacity. These assumption are not only wrong (see, for example, Behnke et al. 1993, Pimbert and Pretty 1995: 5), but simply provide a scapegoat for a problem rather than looking for sustainable solutions. Such a search requires the inclusion of the affected population. The Bedouin need to be part of the project. Their perceptions of the problems, their causes and their possible solutions need to be taken into account. Their needs for their own herds, their access to grazing land, water and supplemental feed need to be considered as well. For without accommodation of their needs, Bedouin will not support the project, rendering the international wildlife reintroduction effort unsustainable in the long term.

The scenario is not as bleak as would first appear since recently a quiet effort in the direction of mobilising community resource management, of encouraging the formation of small 'user' groups, and of building capacity and managing institutional change has begun. For the past two years workshops – initiated by the FAO – have been held at or near the site of the oryx reintroduction project. These have aimed at introducing the concepts of participation into more than just the vocabulary of project personnel. They have brought together government technicians, project personnel, extension teams and the Bedouin. These workshops have been moving, step by step, towards drawing all sides together to work towards a common goal – maintaining the wildlife reserve while at the same time permitting limited resource use by the Bedouin and other inhabitants of the area. The end goal is to achieve further capacity building and truly participatory resource management.

Government efforts to rehabilitate the Syrian desert rangelands in the 1960s initially failed to meet their objectives. Only when the human element was integrated into project development was there some success (Draz 1977). Thirty years on, government and international development agencies again proposed to rehabilitate parts of the desert and to establish a wildlife reserve – without any Bedouin consultation (FAO 1995, Roeder 1996). The lessons learned decades before appeared, briefly, to have been forgotten. Now, however, as a new century dawns, the Syrian government and its international conservation partners are once again looking at the delicate balance which needs to be maintained between pastoralists, conservationists and the environment. Through the medium of participatory resource management, sustainable conservation and development is being sought.

Conclusion

While international conservation and development experts appear to be operating on one level of abstraction based upon imported Western philosophy and technology, the Syrian government seems to have accepted that alternative traditional tribal systems of natural resource use do exist and have some merit. In 1999, for example, a serious drought in Syria resulted in the Minister of Agriculture being pressured by traditional Bedouin leaders and other supporters to lift the ban on Bedouin livestock grazing in all government research stations and plantations in the *Badia*. This pressure was countered by the conservation and international expert group which strongly opposed such a move. It was assumed by the conservation team and other international experts that this would result in a 'free-for-all' and seriously harm the progress that the government research stations had made in the past two decades. The Minister of Agriculture decided to open the government lands to Bedouin herders, in view of the severity of the drought. As Rae (1999) has shown, the Bedouin migrations into the government grazing areas was not an 'open-access' tragedy of the commons. In spite of predictions to the contrary, it followed traditional tribal patterns of natural resource use during drought.

Today, the Bedouin have been pushed back ever deeper into the *Badia*. The Syrian government has taken over to protect the environment from the indigenous population which has lived on it for centuries. It is attempting to revive the 'degraded' *Badia* by reseeding and planting. It has also set about reintroducing large mammal species that have been extinct for half a century, if not more. The imminent failure of these efforts highlights the political nature of environmental protectionism in Syria, which is based on assumptions, mainly of Western origin, that governmental control and authority requires a sedentary population. However what emerges from this study is that a political understanding between government and tribal leadership has supported the continued existence of alternative systems of land use among the Bedouin, which is unofficially tolerated but officially denied. This de facto recognition, it can be argued, points to the philosophical and political bankruptcy of state policy which is supported by convenient but untested 'pseudo' scientific assumptions imported from the West and parts of the former colonial empires of Great Britain and France.

Notes

1   The term *Bedu* or Bedouin means, in Arabic, someone who lives in the *Badia.* It connotes, therefore, a person whose way of life is characterised by raising herds of domesticated animals and moving them about a tribally defined area in search of pasture and water.

2   Land ownership and use rights in Syria were derived from Islamic law. There are four categories of land classification. These are: *mulk,* full private ownership rights, generally granted on land which is cultivated; *miri,* or state land the use of which was allocated to tenants; *matruka,* or public land which was for general use or assigned collectively to a group of settlements; and *mawat,* or dead land which was neither occupied or left for the use of the public. Most of the *Badia* was regarded, from the perspective of central authority as *mawat* land. The Bedouin, however, based their claims to the latter on *urf* or customary law, which recognised varying levels of rights to use and possess the benefits which derived from this land, namely water and grazing. These tribal rights were fluid, evolving and transforming as tribes contested the extent and boundaries of each others' territories or *dirah.*

3   Bedouin animal husbandry is based on risk minimalisation rather than the more common Western market-profit motivation. See Shoup 1990: 200.

4   The Bedouin 'dry-farmed' cereal crops during years of good rain, but the large-scale cultivation in this arid zone had never occurred before.

5   The common Western, urban notion of wilderness as untouched or untamed land has pervaded conservation thinking. Many policies are based on the assumption that such areas can only be maintained without people. They do not recognise the importance of local management and land-use practices in sustaining and protecting biodiversity. Nearly every part of the world has been inhabited and modified by people in the past, and apparent wildernesses have often supported high densities of people (Pimbert and Pretty 1995). In East Africa, for example, the rich Serengeti grassland ecosystem was, in part, maintained by the presence of the Maasai and their cattle (Adams and McShane 1992). There is good evidence from many parts of the world that local people do value, utilise and efficiently manage their environments (Nabhan et al. 1991, Oldfield and Alcorn 1991, Abin 1998, Novellino 1998), as they have done for millennia. These findings suggest, in complete reversal of recent conservation philosophy, that it is when local or indigenous people are excluded that degradation is more likely to occur: 'It suggests that the mythical pristine environment exists only in our imagination' (Pimbert and Pretty: 1995: 3).

6   The term *hema* means to protect or to safeguard. It is said that in early Islamic tradition large swathes of pasture areas and grain fields were set aside as *hema* in order to provide feed for the herds of the Bedouin military units serving in the expansion of the Empire.

7   Today perhaps two thirds of Syria's Bedouin population belong to *hema* cooperatives and associated schemes, although government reports (Al-Sammane 1981) suggest that number is nearly 90 percent.

8   Francoise Debaine, a geographer from the University of Nantes, has conducted a study which compares aerial photographs of the *Badia* from the 1930s with satellite images from the 1990s. Her findings (2000) show that although there has been some degradation in parts of the *Badia* during this sixty-year period, it has not usually been in areas grazed by livestock of the Bedouin.

9   Occasionally the 'primitive' or 'backward' habits of the indigenous people were regarded as attractive for tourism and, in carefully regulated circumstances, a limited number of groups, such as the San in areas of the Kalahari, were allowed to remain in or near traditional lands.

10   The Jordanian Dana Project did not originally integrate the indigenous population into the planning and implementation of the project. In its first few years it relied on a combination of passive participation (limited employment as wardens or guards) and a programme of monetary compensation to buy off the indigenous Bedouin and secure their promise not to use the grazing areas earmarked solely for protected wildlife (Antoine Swene, personal communications). Since then lessons have been learned and a more progressive approach is being applied to other conservation projects in Jordan (Johnson and Abul Hawa 1999). Information on the Saudi oryx and gazelle project at Mahazat As-Said Reserve has been limited to brief public relations information in the IUCN Bulletin (no. 3,1993) and the occasional article in *Oryx* (the official publication of the Royal Society for the Preservation of Flora and Fauna). It is very unlikely that there has been any indigenous pastoralist participation in the planning or implementation of this wildlife reserve which could be regarded as a 'scientific research station' rather than a project aiming at long-term conservation sustainability.

11   During a consultation visit in 1997, I engaged in a discussion about the hiring of local Bedouin for the reserve as a way of beginning to integrate them into the project. The British wildlife expert at the time rebuked my suggestion, saying that 'Bedouin would not work for the salaries I am offering'. The sums concerned were minimal – a matter of US$20 or US$30 a month, less than the standard local wage. The significance of local, indigenous participation for the long-term success of the project, however, seemed to have been lost on the wildlife expert.

6

'ECOCIDE AND GENOCIDE'

Explorations of Environmental Justice in Lakota Sioux Country

Bornali Halder

The Indian plays much the same role in our American society that the Jews played in Germany. Like the miner's canary, the Indian marks the shift from fresh air to poison gas in our political atmosphere, and our treatment of Indians, even more than our treatment of other minorities, marks the rise and fall of our democratic faith – Felix Cohen (in Weaver 1996: 2).[1]

Race, indigenous peoples and environmental justice

Social scientific analyses of the relation between power and environmental resources have proliferated during the past two decades. Research has sought to uncover the connections between political, ideological, social and environmental processes, and has variously framed such connections in terms of class, gender and race (for example, Harvey 1996, Di Chiro 1992, and Bullard 1990, respectively). Concurrent with the burgeoning concern with ecological interrelations is the interest in the processes of globalisation or internationalisation – processes that have transformed the very meaning of local in general (Harvey 1989), and local environmentalisms in particular (Nygren, this volume).[1]

Academics and activists alike have given much attention to the particular relationship between race and environmental issues, or 'environmental racism', particularly in the United States. The term expresses the view that communities and their environments have been systematically underprivileged on the basis of race. Benjamin Chavis Jr. of the United Church of Christ Commission for Racial Justice, for instance, has defined environmental racism as

racial discrimination in environmental policy making. It is racial discrimination in the enforcement of regulations and laws. It is racial discrimination in the deliberate targeting of communities of colour for toxic waste disposal and the siting of polluting industries. It is racial discrimination in the official sanctioning of the life-threatening presence of poisons and pollutants in communities of colour. And it is racial discrimination in the history of excluding people of colour from the mainstream environmental groups, decision making boards, commissions, and regulatory bodies (Chavis Jr. 1993: 3).

The studies of sociologist Robert Bullard are particularly influential in the area of environmental racism. In his classic 1990 study, *Dumping in Dixie*, for example, Bullard identifies race as the key variable in the deliberate siting of waste facilities in five southern communities. He situates his findings by identifying race as an important component in a wide range of environmental practices, for example, the distribution of such municipal services as gas and water supplies, and the layout of urban areas such as highway configurations and commercial development.

Bullard's research has been complemented by other social scientists such as Bunyan Bryant and Paul Mohai (1992), and Laura Pulido (1996). In the United States, the unity of academics and civil rights activists is growing strong and can be seen evidenced at the Conference on Race and the Incidence of Environmental Hazards that was held in 1990 at the University of Michigan, in which several academics and civil rights leaders presented papers.

One of the earliest landmark studies in environmental justice, and more specifically in environmental racism, was carried out in 1987 by the United Church of Christ Commission for Racial Justice. Their report examined the extent of environmental racism in the United States. It found that race is the most significant variable associated with the location of toxic waste sites and that some 50 percent of Asian/Pacific Islanders and Native Americans are living in communities with one or more uncontrolled or abandoned toxic waste sites (United Church of Christ 1987).

A few years later, in 1991, the first national People of Colour Environmental Leadership Summit was held.[2] More than 650 African American, Latino American, Asian American and Native American delegates sought collectively to broaden the parameters of 'mainstream' environmentalism to incorporate multi-ethnic and social justice perspectives. Delegates addressed, among other issues, cultural evidence of environmental racism, ranging from nuclear testing to environmental pollution of water and land by multinational corporations. They contextualised the material within their own cultural traditions, and provided suggestions on how federal departments could reevaluate their policies and practices so that they better promoted environmental justice as it related to peoples of colour. The summit ratified a statement on the Principles of Environmental Justice, which addressed, cultural self-determination, pollution control and sustainable development. Two of the principles focussed on the unique position of indigenous peoples and have a direct bearing on the case-study materials presented in this chapter: '5. Environmental justice affirms the fundamental right to political, economic, cultural and environmental self-determination of all peoples ... 11. Environmental justice must recognise a special legal and natural relationship of Native Peoples to the United States government through treaties, agreements, compacts, and covenants affirming sovereignty and self-determination' (Bryant and Mohai 1992: 215–219).

A growing body of literature examines the particular nature of indigenous rights and its relation to environmental equity (for example, Barsh 1990, Guha 1990, Perrett 1998, Nugent, this volume).[3] The interaction between concepts of self-determination, sovereignty, treaty rights, colonialism and environmental destruction are explored within both academic and indigenous activist contexts. Several writers have focussed attention upon definitions of colonialism in their analyses of indigenous rights and environmental destruction. In 1973, for example, Robert Davis and Mark Zannis (1973: 37) explored the transformation from 'old colonialism', 'with its reliance on territorial conquest and man-power' to 'new colonialism', 'with its reliance on technologically-oriented resource extraction and transportation to the metropolitan centers'. More recently, Jace Weaver (1996: 3) has written that '[e]nvironmental destruction is simply one manifestation of colonialism and racism that have marked Indian/white relations since the arrival of Columbus in 1492' (see also Churchill and LaDuke 1992, Gedicks 1994). Though the language or jargon varies, such debates are frequently expressed in treaty gatherings across Native North America, as well as globally, within the context of the international indigenous rights movement.

This chapter attempts to present some of these issues within the context of particular case materials of the Lakota Sioux, who have traditionally occupied the western region of South Dakota, encompassing the Black Hills, part of the South Dakota Badlands and the Cheyenne River system that drains several Sioux reservations. Although I have had at my disposal a wide-ranging body of material on environmental issues in 'Lakota Country' – principally Missouri River water rights, buffalo and prairie restoration projects, the environmentally destructive effects of tourism in the Black Hills and the Badlands Bombing Range, all of which are dealt with in detail in my doctoral thesis – I have chosen to focus on the politically and culturally important location of the Black Hills. Annexed from Sioux territory by the United States in the late nineteenth century, the Black Hills region has become a symbol of immense political tension. Of the total area of the Black Hills, 1.2 million acres is under the authority of the United States Forest Service; logging and gold mining is carried out there; at one point, uranium was mined in the region; it is also the locus of an on-going land claim on the part of the Lakota.

Lakota elders, political activists and environmentalists have maintained a consistent resistance over the course of the twentieth century against such activities in and around the Black Hills. They have stated that the environmental impacts of such activities – impacts such as water contamination and radiation poisoning – not only constitute a direct attack on sacred land but an attack on Lakota sovereignty. This chapter examines the nature of Lakota discourses on environmental justice – discourses that are intrinsically bound by religious values and political notions of sovereignty and self-determination.

Black Hills land claim

At the time of United States contact with them, the 'Sioux' was a loose confederation made up of seven politically discrete, culturally related groups, one of which was the Teton. Within the Teton group of Sioux were the Lakota, Dakota and Nakota linguistic groups. The Lakota today largely comprise the Oglala, Minneconjou, Hunkpapa and Sicangu bands of Cheyenne River, Rosebud, Pine Ridge and Standing Rock reservations (see Hyde 1937 and 1956).

Lakota Country is a vast ocean of prairie and sky, except for a pine-forested oasis known to the Lakota as *He Sapa*, and known by non-Indians as the Black Hills. Across the sweep of prairie, the Black Hills rise in a volcanic uplift that swelled to life close to a billion years ago. The Hills are composed of granite rock that threads itself throughout the range and binds two worlds together: above, a forest of ponderosa pine, below, an intricate web of caves. Some 79 miles of cave networks have been mapped so far, under a forest that stretches some 120 miles north to south and 50 miles across, covering 6,000 square miles in total.

In the eighteenth century, the Lakota radiated out from the Black Hills to cover a vast region that spanned from the Mississippi to the Rockies, in pursuit of the buffalo. By 1851, a treaty was signed between the United States government and the Sioux that diminished Sioux territory to a tract of land that still centred around the Black Hills, but which now included only the present-day states of Nebraska and South Dakota, areas of Kansas, North Dakota, Montana and Wyoming, and a small portion of Colorado. The Sioux land base was further diminished when the 1868 Fort Laramie Treaty redrew Sioux boundaries to include a portion of present-day South Dakota that lies west of the east bank of the Missouri River, including, just about, the Black Hills. When gold was discovered in the Hills in 1874, the United States government unilaterally annexed the Black Hills from Sioux territory without the stipulated consent of three-quarters of the adult male Sioux population. By the late nineteenth century, the Sioux were confined to small pockets of worthless land that resisted the pull of the plough and civilisation.[4]

Confined to reservations, but undefeated in spirit, the Lakota filed their first suit against the United States for the illegal annexation of the Black Hills in the 1920s. Appeals were entered and thrown out until finally, in 1980, the Supreme Court upheld a Court of Claims award of US$17.1 million compensation, with a 5 percent simple interest calculated annually since 1877. The total compensation package came to US$122.5 million. The Supreme Court rejected the argument that Congress had been acting in good faith and with good intentions in annexing the Black Hills from Sioux territory. The Court found that rations had been used to coerce the Lakota into giving up their land. It recognised that the taking of land had deprived the Lakota of all means of livelihood, and that it had violated the 1868 Treaty. The Court stated that the United States must at last and finally be obligated to make a just compensation to the Lakota. The verdict proved to be the most strongly worded opinion to date in the Lakota case and was scathing in its assessment of United States action. The Sioux resoundingly rejected their right to claim the compensation monies, preferring restitution of treaty lands instead. At the time this chapter was written, it was estimated that all award monies owed to but untouched by the Lakota had risen in value to around US$400 million.[5]

In recent years, several members of Congress have sponsored bills on behalf of the Lakota and the land claim. All bills have failed to pass.[6] During the hearings for one of these bills, members of the Lakota Nation argued that return of the Black Hills was a matter of legal and spiritual right. Of the latter, the return would enable the Lakota to exercise their religion freely – so integral are the Black Hills to ceremony. They also argued on the basis that a strong, religious kinship to the Hills had existed for the people since the beginning of time, when they emerged out of a cave in the form of buffalo. Various sacred sites in the Hills were named, and testimony supplied, linking the Hills to a complex system of star theology. Other arguments advanced included the assertion that return of a portion of the Hills would enable tribes to develop their economies in a self-sufficient way and to strengthen religious identity. One Oglala testifier said: 'I feel very strongly that with the return of the Black Hills, the bond of imprisonment would be broken. By that happening, many young Indian people, many adult Indian people, and many elderly people once again can start having a better perspective of who they are as Lakota people'.[7]

Such cultural beliefs have been expressed for at least the past 120 years. Physician and anthropologist James R. Walker collected several myths at the turn of the twentieth century expressing the importance of this location. Such myths described how the 'soup of life'[8] began there: the beginning of space and time, the cyclical motion of life itself set in motion by the mythological figure of *Inyan*, the Rock, whose remains saturate the ancient granite core of the Hills (see Walker 1917). The Lakota, as the *Pte Oyate* or Buffalo Nation, lived under the earth in a network of caves. They emerged to the surface through an opening now concealed within federal land: Wind Cave National Park. The Lakota were born and they died in the Hills: it was a major burial site for them, along with the Badlands to the east. The Black Hills are the home of the revered and feared *Wakinyan*, the Thunderbeing or Thunderbird. Ceremonies, such as vision-quests or *hanbleceya*, sweats or *inipis*, and sundances were and are still performed there. Tobacco ties, plugs, and even cigarette packets, strips of cloth in the sacred colours of blue, green, red, yellow, black and white, and beds of sage and braids of sweetgrass can be found throughout the region, especially around sacred sites such as Devils Tower and Bear Butte.[9] On a utilitarian level, the Hills were important because they had once – before the 1877 annexation, at least – served as a major source of food, water and shelter for the Lakota, especially during the winter months. One Oglala Lakota elder described it as a 'refrigerator' and told me that the Hills were given to the Lakota people by *Wakantanka*, the Great Mystery or Great Spirit, to take care of and to survive from. He said: 'The old people, the old people a long time ago, called [the Black Hills] the purse, the mother's purse. Because it has everything in it. It had all kinds of jewellery and all kinds of gold. But that's for mother earth. That's part of the earth. It's the earth. And then, the Indian people go over there and they hunt, and they gather, in the summer time they gathered. It's a rich piece of land' (Field notes, March 1998).

The Lakota followed the buffalo – their primary food source – into the Hills. The buffalo entered the Black Hills through a gap now known as Buffalo Gap, especially during the winter, where, like the Lakota, they retreated to seek shelter from the punishing Plains blizzards. Wild fruits, vegetables, water and meat were not the only provisions obtained from the Hills to furnish the Lakota's survival. The Lakota also obtained their lodge poles to construct their tipis from the Black Hills.

Today, two treaty councils in Lakota Country are concerned with the Black Hills land claim: the Teton Sioux Nation Treaty Council and the Black Hills Sioux Nation Treaty Council. The councils are composed largely of elders. Much of the activist work on the Lakota Sioux reservations is handled by the elders. One of the concerns expressed to me by elders was the perceived lack of active involvement in treaty issues by the younger generation. Indeed, virtually every treaty meeting I attended in 1998 had few under-fifties in attendance. This is not to say that the younger generations do not care about the Black Hills land claim and other treaty issues – most whom I spoke with are concerned – but it is the elders who are able to devote the most time to the cause. Another reason expressed to me was the fact that traditionally in Lakota culture the elders have always taken the lead in political issues, and younger Lakota defer to their authority and wisdom in such matters.

The two treaty councils hold regular meetings each year on most of the Lakota and Dakota reservations in South Dakota. Through pursuing the land claim, the councils have developed and maintained networks with such national and international organisations as the Sierra Club, the Indigenous Women's Network, the Indigenous Environmental Network and the United Nations Working Group on Indigenous Populations. The two councils situate the claim within a decidedly international context of worldwide indigenous struggles and maintain strong links with and interests in the activities of other indigenous struggles. In addition to the land claim, the Black Hills Council concerns itself with all national matters pertaining to treaty boundaries – for example, energy development, environmental problems, water rights and reservation land issues.

The identification of land reclamation with sovereignty and self-determination has been a long-standing one for the Lakota. However, the Lakota employment of the language of colonialism, decolonisation, human rights and indigenous rights, and the identification of themselves and their case with other indigenous populations fighting for land reclamation, environmental protection and political independence around the world, appears to be a more recent phenomenon – building and developing most intensely over the past thirty years. It can be assumed that the increasing contextual internationalisation of the Black Hills land claim has arisen due to the nurturing of national and international indigenous networks.

The Teton Council is perhaps the most international of the two treaty organisations. It was first formed in 1894 to ensure the implementation of the 1868 Fort Laramie Treaty. For the last fifteen to twenty years the council has sent elders and other representatives to the United Nations (UN) to argue the land claim before such working bodies as the Working Group on Indigenous Populations, and the Subcommission on Prevention of Discrimination and Protection of Minorities. It has also been preparing documents for submission at the International Court of Justice in the Hague.

The goal of the council is that the Lakota Nation be recognised as a sovereign nation on sovereign land. Sovereign territory centres around the Black Hills. The Council describes the United States as an imperialistic power that has sought to undermine, diminish and abolish not only the Lakota land-base but Lakota rights to self-sufficiency and self-determination. The council accuses the federal government of chemical warfare, disease and poverty. It accuses tribal government of being a tool of the federal government and thus not concerned with the affairs and rights of the traditional people who, they say, are cheated time and time again. The only government the council acknowledges are the traditional political groupings or governments of *tiyospaye*.

An Intervention of a Dakota elder at the UN in 1998 gives an idea as to the concerns of and language used by Lakota treaty councils today:

> In our homeland, the coloniser has been severely limiting our right to development. Before the coloniser came to our homeland, we did not have to refer to ourselves as 'indigenous'. We are and were Lakota, Dakota and Nakota, the Allies,

living on the land for which we are responsible. We live by the Natural Law given to us by the Creator with instructions for the care of our people, our land, our culture and our natural world. From this Natural Law comes a way of life so beautiful it requires very little improvement. It is this Law that guides us in determining our systems for health care, shelter, economic and social programs and the institutions by which we are governed and live. When the colonisers came into our territories, we gladly shared our way of life with them. A cornerstone of our Natural Law is generosity and caring for those weaker than you ... Soon we learned though that the colonisers came with the intention of taking our land and the natural wealth found on, above and under our territory. They take from us our medicinal plants, the gold and minerals inside the ground, and the water flowing over our territory. Our land was taken without our consent and treaties were signed and immediately broken when the conditions did not satisfy the coloniser's greed. Every inch of territory upon which the coloniser lives has been taken at gunpoint (Grey Owl 1998).

Natural Law, a recurring concept in environmental protection and land reclamation meetings on Pine Ridge and Cheyenne River reservations, was described by one Lakota man as such: 'We had and still have a way of life so beautiful that we never needed drugs or alcohol because we had the connection to our land. Our Creator gave us this land to use and take care of and he gave us ways to live with it. This is what we call the Natural Law' (Field notes, February 1998).

Mining for gold

Mining [has] taken away the beauty of this country, the Black Hills. You know, God put it there for us to take care of, and we could eat from that place: there were a lot of wild fruits and animals and deer and buffaloes. But when you go through mining or clear-cutting, that's a lot of species' habitats ... We like to live with the environment, like good environment. You know, we could go out and hear the birds, and clean air, that's what the people, way back, ... our ancestors, they were pretty healthy. We lived off the ground, off the land (Oglala Lakota elder, March 1998, Pine Ridge Reservation).[10]

Much of the Black Hills today is under the supervision and 'protection' of the United States Forest Service. Mandated by the Multiple Use-Sustained Yield Act (1960), the Forest Service employs a multiple-use philosophy that seeks to balance recreation, logging, grazing and mining with environmental preservation. This multiple-use concept has caused multiple conflicts, as we shall now see.

The earliest recorded discoveries of mineral resources in the Missouri River West area – defined as 'Sioux Country' under the 1851 Laramie Treaty – were made by explorers in 1875, who alerted the general public to deposits of gold in the northwestern Black Hills. A few years later, copper was found along the North Platte River, now in Wyoming, and iron prospecting began in 1888. Coal discoveries were first reported in 1877, just northwest of the Black Hills. By 1877 the Sioux had lost their rights to the entire Black Hills region. By 1928 the gross value of coal production in the Missouri River West area was reported at US$31 million. Most of the coal found in the Wyoming part of the Black Hills was used by railroads to power their coal-fired steam engines. Other minerals that have since been discovered and mined in the region have included mica, silver, tin, tungsten, quartz, iron and petroleum (Shenon and Full 1969).

It has been gold, however, that has proven to be the area's greatest lure and has accounted for around two-thirds of the total mineral production in the Black Hills. Although numerous reports regarding the existence of gold in the Black Hills had been floating around along the 'moccasin grapevine', it was Lieutenant G.K. Warren who supplied the first official statement of its existence after he commanded an expedition of western Dakota in 1857. Further official reports followed, but it was not until the government sent Lieutenant Colonel G.A. Custer on an expedition through the Hills in 1874, to investigate communication routes and possible military posts and which found gold, that the rush to the Black Hills began in earnest.

Immediately after the cession of the Black Hills from the Sioux, the counties of Lawrence, Pennington and Custer were formally organised and the mining for gold became official. By 1876 the non-Indian population of the Black Hills was estimated at 10,000.

In 1876 the Californian-owned Homestake Mine was located in the vicinity of the towns of Lead and Deadwood. The Homestake belt in the Hills has proven to be the most successful gold mining region in the United States and, for several years, in the world. As part of their work for the on-going *Sioux Nation vs. United States of America* court case, Shenon and Full (1969) appraised the '1877 fair market value' of the Homestake gold belt to be US$9,812,175. Their appraised '1877 fair market value' for gold found throughout the Black Hills was US$13,558,489. Between 1878 and 1962 gold bullion production at Homestake's Black Hills mine totalled US$715 million.

The process of mining Homestake has largely been open-pit, heat-leach gold-mining where the ore is blasted out of the mountain, hauled in trucks, crushed, and then a cyanide solution used to separate the gold. The system has led individuals and organisations, including nearby Lakota tribes to make accusations of water and air contamination. Several suits have been filed against the company by tribal, state and federal governments and by local environmental organisations.

In November 1997 the Cheyenne River Sioux Tribe, with the United States Justice Department, filed a civil suit against Homestake Mining Company for allegedly dumping thirty million tonnes of toxic mine tailings, including cyanide, mercury and arsenic, into waterways of the Black Hills – such as Whitewood Creek and the Cheyenne and Belle Fourche Rivers – which drain the Cheyenne River Reservation. In addition to the tailings, the suit also alleges daily discharges of zinc, copper, cadmium, chromium, lead, nickel and selenium. The suit is seeking monetary compensation to cover the one-hundred-year period the mining company has been operating in the Hills. One of the lawyers acting on behalf of the tribe told me that the suit was a critical one in terms of protecting the health and safety of Lakota residents, of protecting the environment that is being claimed by the Lakota and of protecting Lakota treaty rights (Field notes, August 1998). At the time of writing, the case was still in court.[11]

Uranium mining

Discoveries of uranium ore in the southern Black Hills were first announced to the public in 1896. The next time it was brought to public attention was in 1951, when more ore was located nearby. Coming just after the Second World War, the discovery caused much excitement among the local populations. Residents provided much of the labour for the mining companies that moved in and began to profit from the deposits. Little concern was expressed over the fact that the mining was taking place in the vicinity of ancient Indian pictographs, petroglyphs, flint quarries and burial sites, and little was understood about the risks to health, land and water from radiation pollution.

By 1980 around 2,345 square miles of the Black Hills was under uranium mining leases. In this area, 5,748 uranium claims were held, mostly in the National Forest and 355 square miles of leased land. The environmental effects of such activity have been disastrous. In 1962, for example, an estimated two hundred tonnes of uranium mill tailings washed into a tributary of the Cheyenne River, the primary source of surface water for Pine Ridge Reservation. The United States Environmental Protection Agency began expressing concerns about the levels of pollution around the southern Hills, and a group calling itself the Black Hills Alliance (BHA) was formed in the mid- to late 1970s.[12]

A mixed group of Lakota and non-Lakota activists, the BHA declared the region a National Sacrifice Area and began to agitate and educate local people about the threats posed to the Hills by uranium mining. Citing figures issued by various governmental and private sources, they drew attention to a sharp increase in the rate of cancer in the county most affected by uranium mining (Fall River County),[13] and demonstrated increased levels of radioactive pollution of aquifers that supplied the major water supplies to Pine Ridge Reservation and nearby non-Indian communities bordering the reservation. The BHA also chronicled and monitored the progress of the various multinational corporations that had uranium interests in the Hills – companies such as Union Carbide Corporation, Tennessee Valley Authority and Gulf Oil, which, with permission from the Forest Service, had been drilling for the ore in the Hills since 1975.

Pine Ridge Reservation, on the southeastern cusp of the Black Hills, was particularly hit by energy development and, some have argued, singled out in a form of 'environmental racism'. In 1979 BHA announced that Union Carbide had received funds from the United States Department of Energy to explore one-fourth of the reservation for uranium. The Oglala tribal council refused to authorise any permits and the plan was quashed. In 1980, Women of All Red Nations (WARN), an alliance of female Lakota activists, released a report that showed a statistical correlation between high incidences of spontaneous abortions, cancer and birth defects, as well as polluted and radioactive water contamination on Pine Ridge (Women of All Red Nations 1980: 1–9). Water contamination was not only associated with uranium mining in the Hills, but also with activities on the Badlands Gunnery Range during and after the Second World War,[14] and chemical herbicide and insecticide run-off from off-reservation farming activities. WARN cited a report by a Rapid City biochemist who found that Pine Ridge water contained 'lethal' dosages of radioactive particles: nineteen picocuries[15] of uranium radiation per litre in surface water from subsidiaries of the White River, which flows into and through the reservation, and fifteen picocuries per litre in groundwater in the Lakota Aquifer under Red Shirt Table community – which also happened to be the closest community to the Gunnery Range (ibid.: 5). The report also highlighted the high levels of nitrates in reservation water samples, and points the blame in the direction of gun blasts carried out on the bombing range (ibid.: 6). To underscore their grim findings, WARN also described how '[c]hildren swimming in subsidiaries of the Cheyenne River have frequently been admitted to the hospital with body sores' and doctors could not determine the cause (ibid.: 6).

In 1980 hundreds of Lakota and non-Lakota people congregated at an International Survival Gathering in the Black Hills, at which sacred land, environmental protection and indigenous peoples' treaties and treaty rights were affirmed. The Lakota activist, Madonna Gilbert Thunderhawk concluded the proceedings thus: 'The land is not dead yet. There is a struggle that will go on ... We will go home and get back into the long, hard, tedious process of education, agitating, organizing. This is the way we will save our Mother, the Earth' (Black Hills Alliance 1980: 2).

After just a few years of active pressure, which included taking companies to court, the BHA proclaimed their work complete and disbanded. To date there are no uranium activities in the Black Hills. Parts of the southern Hills, which bore the brunt of uranium mining, have undergone extensive clean-up operations.

Energy development still poses a threat to the Hills region, however. The Black Hills lies on the eastern edge of what has been called 'the great coal basin of the nation'. Coal is found in abundance in Wyoming. In 1971 a group calling itself the Power Supply Entities of the North Central and

Rocky Mountain Region released a report that called for the construction of some twenty-two power plants and six gasification plants in the area (United States Department of the Interior 1971: 11). In addition to these, seventy-three power plants were planned by other companies. These coal-fired power plants are slowly being built. One has been completed in Gillette, Wyoming, one in Wheatland and another in Montana. Because uranium is present in these coal beds, there is local concern that the burning of coal is releasing low-level radiation. Like any type of mining, coal mining uses up great quantities of water and there is concern that water is being contaminated. The Big Horn Mountain region, where much of the coal resides, supplies the major water supply to the aquifers that supply water specifically to Pine Ridge Reservation. Moreover, the water moves through the Hills, which acts as a recharger of all the water that comes out of the Missouri in the east and the Big Horns in the west.

In 1976 and 1977, the South Dakota State Geologist declared several parts of western South Dakota 'ideal' for a nuclear power plant and nuclear waste disposal (Sixth District Council of Local Governments 1976: 8, 1976: 12). The same report notified the public that the Union Carbide Corporation had been awarded a five-year grant by the Energy Research and Development Administration (ERDA) to explore western South Dakota for nuclear development (ibid. 1976: 6). Indeed, during the Second World War, one small town in the southern Hills – Igloo – was an atomic bomb storage area that was not closed down until the mid-1950s. A train used to run between Ellsworth Airforce Base in the northern Hills to Igloo, transporting nuclear bombs. Today, Igloo has been declared a Superfund Clean-up Site by the Environmental Protection Agency and the town, which once housed some 10,000 residents, is now virtually deserted.

Shortly before the Gulf War, the United States government wanted to use one of the Lakota's seven sacred sites in the Black Hills – Hell Canyon – as a weapons testing area for uranium-tipped shells that were subsequently to be used in the war. Once again, public outcry, fuelled by a coalition of Lakota and non-Lakota activists, put a stop to the plans. Not only is the area a sacred site, but it harbours a wild horse sanctuary as well as pre-historic petroglyphs and tipi-rings. During my fieldwork the entire area was under heavy road construction to facilitate recreational tourism, a different kind of threat to the Black Hills.

Military interest in the mineral composition of the Black Hills still remains. The mineral tantalum is still mined for its use in bullet-proofing military aircraft windshields.

Today, plans are afoot for the building of a railroad to transport coal from these coal beds in Wyoming, around the southern Hills, across the Cheyenne River border of Pine Ridge Reservation, and back east.[16] At the time of my fieldwork, several organisations, including the Black Hills Sioux

Nation Treaty Council and the Lakota Landowners' Association, and local chapters of the Sierra Club and Peace and Justice Group, were campaigning against the Dakota, Minnesota and Eastern (DM&E) railroad plans. Environmental concerns expressed included the passing of the railroad through several areas the Sierra Club have requested the Forest Service to designate wilderness areas, the threat of coal spillage and pollution along the Cheyenne River, air pollution from the coal dust and train smoke, fire threats to parched summer grasslands, the noise and disruption to local residents and wildlife, and the possibility that the railroad might also be used for nuclear and other waste transportation. One Lakota woman even posed the question: 'Is it possible that there are ideas of talking the Oglala Sioux tribal council into opening up a major waste dump site, hence the nearness [of the proposed railroad] to the reservation border?' (White Face 1998: A4). She went on to situate the issue within the Lakota concept of Seven Generations, arguing that the long-term consequences for future generations of the environmentally damaging effects of the line had not been considered: 'That old Lakota philosophy of thinking ahead for seven generations was not such a bad idea … Personally, I think there should be an international mandate that everyone in the world is required to follow that says all development of resources must first pass the test of the effect on the seventh generation' (ibid.).

Lakota concerns surrounding the DM&E railroad revolved around treaty issues. At several meetings at the Oglala tribal offices at Pine Ridge, one Oglala Lakota elder – referring to the Union Pacific Railroad, which, in 1862, failed to live up to its promise of Sioux compensation – exclaimed: 'They haven't paid us for the *last* railroad they built through our country, why are they thinking about doing another one here?' The entire region that the railroad will pass through is claimed by the Lakota under the terms of the 1868 Fort Laramie Treaty. Referring to railroad companies, an Oglala Lakota member of the Black Hills Sioux Nation Treaty Council, said to me: 'How long are they going to go on and on violating our treaty rights? We're not going to shut up and just turn over!' (Field notes, October 1998).

In 1998, this same man recounted to me what he told a group of DM&E representatives:

> If we were going to build an Indian railroad right across the Pope's front lawn, I said, what do you think the world would say? All the Christian people, the Catholics, they would be jumping mad because we'd be violating sacred ground, right through the Pope's holy city. So I said, I could probably come that close to a comparison. And I think after last week some of them understood it. But of course, they're working for a corporation, they have no choice but to continue to hammer away at the Indians to build this railroad. The lady … who was representing DM&E … said, 'If you could show us a piece of sacred ground … we will more or less fence it off and we will leave that part alone.' I always thought our conception of sacredness doesn't have four walls. It doesn't have a perimeter. It doesn't have a boundary. It's out there (ibid.).

He told the representatives that the Black Hills was the location of Lakota warrior and leader Crazy Horse's birth, although precisely where, no one knows:

> So the next question that was brought up was, 'Well, if you could identify exactly where Crazy Horse was born, we'll ... put up a boundary.' I said, 'No, you can't do that! That's his birthright, and if that's the way you're thinking, if you're thinking *boundary*, he fought for a whole 1851 Treaty boundary, and that's three states wide! That was his territory. That's what he defended. And you *brought* Custer over.' So again they backed off. So we'll be going out there, looking for Crazy Horse's birth site. Once we establish some of this stuff, we have to [pauses]. It's getting to a point where we have to define sacred space now (ibid.).

Defining sacred space for the Lakota, however, is no simple thing:

> You see, in Indian terms we understand what we're talking about. But in the English term, there's no such thing... [S]acred space is without walls, without containment, without a perimeter, or without lines – it's there. That's the way we see it. It's like the Black Hills, that's a sacred ground. Again, they say, 'How far? It could be all the way up, all the way down' (ibid.).

At the time of writing, the DM&E were still considering the route as a possibility.

Lakota environmental values

Much has been made in recent years of the tendency to generalise about indigenous environmental values (for example, Ellen 1986, Milton 1996). Certainly, the data presented in this chapter may have given the impression that in matters of land reclamation and protection, the Lakota Sioux speak with a uniform voice. Of course, reservations are as fractured as all other communities. Often reservations are unofficially split between 'traditionals', who tend to favour environmental preservation, and 'non-traditionals', who tend to see development as Native Americans' only hope of economic and cultural survival. This appears especially true of the Oglala Lakota of Pine Ridge Reservation, wherein traditionals tend to view non-traditionals as tools of the federal government who relinquish traditional values, rights and land for economic and political gain, even though many such traditionalists participate in the tribal government. Some tribal governments across the United States – such as the Navajo, who leased mining rights on their reservation out of economic necessity – have sold off their resources, sometimes without the majority consent of their people. Others have actively fought alongside environmentalists against destructive corporate activity on reservation and treaty lands.

During the course of my fieldwork I rarely heard a negative Lakota response to the work of the elders and activists in fighting for land restitution and protection. Numerous reasons may be advanced for this, some of which lie in the limitations of my fieldwork: some people felt inhibited to talk freely with me because I was an outsider, and because the object of my fieldwork was to gather data on Lakota environmentalism, much of my time was spent with elders, activists, spiritual practitioners and other traditionals – the primary groups involved with environmental and treaty issues. The few negative remarks shared with me related largely to the land claim rather than environmental protection and expressed the view that the land claim was futile, that the Lakota would never be able to reclaim the Black Hills, and that the money spent fighting for land reclamation would be better spent on economic and social necessities.

Literature on Native Americans has tended to posit two extreme views on their environmental values and practices. One view may be termed the 'harmonious model'. It depicts Native Americans as lovers of nature, living in unceasing harmony with it, whose subsistence activities have caused no or minimal changes to the North American natural environment. The 'destructive model' presents Native Americans as exploitative despoilers who have ravaged the natural world and changed it irrevocably (see Gill 1987, Krech III 1999, Martin 1978).

'Traditional' values of spiritual and cultural connectedness to the natural world were strenuously affirmed at every meeting I attended and in every formal and informal conversation I had during my fieldwork year. Certainly, exposure to United States education and media, the increasing tendency to work with non-indigenous environmental organisations, and the nurture of national and international indigenous networks have undoubtedly increased Lakota utilisation of a 'mainstream' environmental vocabulary. 'Stewardship', 'conservation', 'preservation' and 'holism' are frequent refrains within Lakota environmental discourses. Environmental values flow in many directions, however, and several activists were clear to point out to me that such concepts may themselves have been influenced by those rooted in their own tradition – concepts such as that expressed in the phrase *mitaku'oyasin*, all my relations, which is a frequent refrain in Lakota ceremony. It expresses the belief that all things – visible and invisible, human and nonhuman – are related as spiritual kin and bound together by the binding force that is *Wakantanka*, commonly glossed in English as the Great Spirit or Mystery (see Powers 1975, Walker 1980).

Indigenous rights and environmentalism

It could be said that the Lakota first suffered at the hands of environmental racism back in the nineteenth century, when the treaties of Fort Laramie successively reduced their landholdings, thereby dispossessing them of their means of spiritual, historical and economic sustenance. Environmental injustice can be said to continue today as the Lakota are excluded from environmental, corporate and federal decisions regarding the Black Hills land claim area and as their ceded and unceded lands are subjected to a new form of colonialism – resource colonisation.

Lakota Sioux movements for treaty rights and the protection of the environment intersect and have done so since the Lakota Nation first began agitating on behalf of their land and water at the turn of the twentieth century. Meetings, gatherings and conversations relating to environmental protection always occur within the context of treaty rights – rights which include land reclamation, self-determination and cultural renewal.

Although the roots of environmental justice reach further back in Western, non-indigenous history (Gottlieb 1993), the non-indigenous environmental movement finally woke up to the inseparability of social justice and environmental destruction when the environmental justice movement emerged out of the civil rights actions of the 1960s and 1970s. By recognising the environmental rights of communities of colour, the roles race and socio-economic status play in the distribution of environmental risk, and the importance of indigenous peoples' political and sovereign right to own, manage and demand protection for indigenous land, the environmental justice movement has injected new life into the discourses and practices of non-indigenous environmentalism over the past two decades (see United Church of Christ 1987, Wenz 1988, Bullard 1990, Bryant and Mohai 1992).

It is all too easy to generalise the ideological positions of 'mainstream' environmental organisations. Norton (1991), for example, identifies at least seven 'mainstream' and Western environmentalisms: Judeo-Christian stewardship, deep ecology, transcendentalism, constrained economics, scientific naturalism, ecofeminism and pluralism/pragmatism. Generally, however, one can say that such mainstream discourses and practices have been predicated upon the ideological separation of nature and culture. The consequences of such dualistic thinking have frequently meant that such 'Group of Ten' organisations as Friends of the Earth, the Audubon Society and the Sierra Club campaign for wilderness preserves or ecological zones – preserves that exclude all humans but, perhaps, conservation managers and ecotourists. Indigenous peoples have found themselves alienated from a rhetoric of wilderness preservation and environmental protection that excludes them from their own treaty lands. For instance, one environmental activist and educator on the Cheyenne River Reservation, familiar with both a mainstream environmental discourse and Lakota cultural traditions, identified for me what he determined to be fundamental differences between Lakota and Euro-American environmental perspectives:

> Dualism is like the European way of understanding the way the world works, and that is that nature is in this larger circle and man is a smaller circle, but he's outside of that, managing it from afar. But the opposite of that is holism, and that describes the Lakota and pretty much Native perspective, where man is a smaller

circle, just one of many smaller circles inside that bigger circle of life. And they're a *part* of that, they're not *apart* from it. The larger circle needs all of those integral parts inside in order to be whole. Whereas man is not a separate function from the rest of it. You see that in the way that the environmental organisations in this country view how they want to *fix* environmental problems ... When they want to restore land to its former pristine state, what they will do is they will buy up a certain piece, or they'll have a certain piece, and they will try and make everything whole inside of there, but there will also be a boundary around it that says, '*Humans* can't go in there,' or, 'Humans aren't part of this, so stay out!' That's not so with us. The opposite is true. That if we're going to restore this piece of land then the Native peoples have to be a part of it as well (Field notes, August 1998).

Certainly in western South Dakota, attempts are being made by the local environmental movement to address the social justice needs of the Lakota people. Environmental action in the treaty lands of the Black Hills region has involved a partnership of both Lakota and non-Lakota activists, and meetings of local chapters of the Sierra Club and Peace and Justice not only acknowledge the treaty rights of the Lakota, but regularly invite elders and other treaty activists to present the Lakota case. Across the United States, activists in the environmental justice movement are championing a reconceptualisation of environmentalism that incorporates the needs of indigenous peoples (Gedicks 1993).

Notes

1   This chapter is based upon twelve months of doctoral field research at Pine Ridge and Cheyenne River Reservations and in Rapid City in South Dakota between 1997 and 1998. Field data were gathered by means of formal and informal interviews and conversations with activists, elders, spiritual leaders and with those who did not define themselves as any of the above. The author also attended countless treaty gatherings on the reservations, in which Lakota people testified – in Lakota and in English, and to a predominantly Lakota audience – on a variety of political and cultural issues. Data were supplemented by published and unpublished written documents, such as newspapers and private papers and letters.

2   First National People of Colour Environmental Leadership Summit. 24-27th October 1991. Washington DC.

3   The term 'indigenous peoples' is not clearly defined in international law. In its broadest sense, 'indigenous' may apply to a particular group of peoples who have inhabited an area for an extended period of time. Colchester writes that, more specifically 'the term "Indigenous Peoples" has gained currency, internationally, to refer ... to long-resident peoples, with strong customary ties to their lands, that are dominated by other elements of the national society' (1999: 5). In addition, Article 8 of the United Nations Draft Declaration on the Rights of Indigenous Peoples recognises the principle of self-identification, whereby groups of people have the right to define themselves as indigenous (UN World Council of Indiginous Peoples 1994).

4   1851 *Treaty of Fort Laramie* (11 Stat. 749), 1868 *Treaty of Fort Laramie* (15 Stat. 635), and Act
    of February 28th, 1877 (19 Stat. 254).

5   *Sioux Tribe v. United States*, 318 United States 789 (1943), *Sioux Tribe v. United States*, 2 Ind.
    Cl. Comm. (1956), *Sioux Tribe v. United States*, 146 F. Supp. 229, *Sioux Nation v. United States*,
    33 Ind. Cl. Comm. 151 (1974), *Sioux Nation v. United States*, 220 Ct. Cl. 442, 601 F. 2d. 1157
    (1975), 448 United States 371 (1980).

6   Sioux Nation Black Hills Act (S.705, introduced 10 March 1987), Sioux Nations Black
    Hills Restoration Act (HR 5680, introduced 19 September 1990).

7   S. Hrg. 99-844, *Sioux Nation Black Hills Act: Hearing before the Select Committee on Indian Affairs,
    United States Senate, Ninety-ninth Congress, Second Session, on s. 1453*, United States Govern-
    ment Printing Office, 1986.

8   A phrase used by one of my research participants (October 1998, Pine Ridge Reservation).

9   During my fieldwork, sundances were held at Bear Butte, Devils Tower and Wind Cave.

10 Major sources of information for this section came from ex-Black Hills Alliance members,
    members of the Black Hills Sierra Club, and various meetings of the South Dakota Peace
    and Justice group, the Teton Sioux Nation Treaty Council and the Black Hills Sioux Nation
    Treaty Council. The staff at Rapid City Library also supplied me with useful documents and
    statistics.

11 In 1982 the Oglala tribe sued Homestake for US$6 billion for trespassing and illegal gold
    extraction. The suit was thrown out of court.

12 Figures in this section from documents of the Black Hills Alliance were given to me from
    a private collection.

13 Several Lakota men worked the Edgemont Mine, in the southern Hills, in the 1950s. One
    such man, now an elder living in the Hills, told me how all his ex-colleagues who had
    worked with him seemed to be dying of cancer now. Few of his other friends – those who
    had *not* worked in the uranium mine – seemed to be dying, he said. The correlation,
    though not explicitly stated by him, was implied.

14 In 1942 the federal government acquired land in the northern portion of Pine Ridge
    Reservation. The Badlands Gunnery Range, as it was known, was used extensively from
    1942 through 1945 as an air-to-air and air-to-ground bombing site. It was also used for pre-
    cision and demolition bombing exercises. After the Second World War the site was also
    used by the South Dakota National Guard as an artillery range. Today, 2,486.40 acres of the
    former Gunnery Range is retained by the Air Force with active status.

15 Picocurie is the measurement for radiation. Any amount above five picocuries per litre is
    considered dangerous by the United States Public Health Service.

16 Around forty trains with 120 cars of coal a day will trundle along the Pine Ridge Reserva-
    tion, close to the Oglala community of Red Shirt, which has already been impacted by the
    Badlands Bombing Range.

7

PROMOTING CONSUMPTION IN THE RAINFOREST
Global Conservation in Papua New Guinea

David M. Ellis

Introduction

In this chapter, consumption is employed as a window through which to consider inequalities of scale and power in resource use and conservation. Practices of Pawaia people in the Pio-Tura region of Papua New Guinea are compared with those of biologists working for a project to conserve biodiversity on their lands.[1] Project objectives, and the implications these might have for local and global consumption, are also discussed. The history of the paradigm of 'conservation and development' is considered within the context of consumption, as are the ramifications of trading in carbon credits. The case presented highlights the imbalances between different regimes of consumption: expatriate biologists tend to be reliant on high levels of consumption while local people tend to be self-sufficient. At another level, the rainforest becomes a marketable commodity which can be purchased in order to bolster economic growth and consumption elsewhere. Certain activities encouraged by biologists serve to promote consumption *in situ* in the rainforest. At the same time, the economic expansion of countries eager to trade in carbon credits to avoid cuts in their own carbon emissions promotes consumption in the name of saving the rainforest.

The chapter also raises a number of questions concerning ethnographies of environmentalism: Why should we undertake comparative ethnography of local people and conservation practitioners? Why should we juxtapose descriptions of ethnographic encounters with these people against an overview of globally legitimated environmental policy? What ironies are revealed in this process and why are they significant? These questions will be revisited in the conclusion. What follows is a meditation on the contradictions of global ecological 'management' and also of being an ethnographer in an era of environmentalised development jargon. An analysis of consumption practices is a means of approaching these issues rather than an end in itself.

Absurd connections?

In March 1997 I left Haia, a village on the lands of Pawaia people near the intersection of Simbu, Gulf and Eastern Highlands provinces in Papua New Guinea, after my first three months of living there. I took a small plane from the grass airstrip amid the mountainous forests of the Pio-Tura region to the highlands town of Goroka. I flew to the capital Port Moresby three days later, during the height of a crisis concerning the government's plans to hire mercenaries to impose peace on the island of Bougainville.[2] There was rioting and looting in the town, and people were concerned about the potential for a military coup. The centre of town was under siege and the entire city was in a state of chaos for a few days.

I needed to meet with one of the expatriate advisers to the Papua New Guinean non-governmental organisation (NGO) working with people in Haia and the Pio-Tura region for the conservation of biodiversity on their lands. As the office in the centre of town had been closed due to the rioting and looting, we arranged to meet at a safer location. I went round to his apartment on a Saturday afternoon.

On that hillside, in a fortified compound guarded by security personnel, life seemed far removed from the tension and rioting which had characterised the previous days. For me, it was one of many moments in a succession of situations of extreme contrast. A few days before, I had been living in a dwelling made from bush materials in a clearing in an expanse of rainforest. Arriving in town again was bewildering in itself. Being in a city under siege felt surreal. From a perspective within the forest, what would it matter if there was a military coup here? If there had been no batteries to operate a radio, we might not have heard about it for weeks. How far away seemed the arena of public life with its politicking and corruption, its daily struggles for survival.

The compound resembled a holiday village. White children were playing with inflatable toys in an outdoor pool. Arriving there was a further step along a continuum of absurdity in terms of the differentials of wealth, power and livelihood which the so-called 'global' condition presents. So this is where practitioners of community-based development and conservation live, I reflected. Fenced in from the outside world, in concrete apartment blocks designed as worlds unto themselves, in the Western image.

The apartment was high above the town. It was bright, airy, brilliantly white. Its atmosphere was like that of a minimalist office, perhaps in New

York or Frankfurt. There was steel and leather furniture. The walls were adorned with exhibits of indigenous artwork, reflecting the trajectory of the inhabitants' careers in global conservation and development.

I was invited to have a drink. We installed ourselves on the balcony and had a lengthy discussion. I looked far out across the sun-burned hills and down to the harbour, wondering where we really were, whose reality the riots had been the previous day, whether the forest existence had been a part of my world and whether this controlled environment in which I found myself was a world I could inhabit.

This extract from my fieldnotes conveys the significance of the moment:

> He is sitting, composed, in his environment. We have had a long and at times heated discussion about the differences in approach between conservation biology and social science. With one hand he holds a polystyrene receptacle which keeps his can of Coca-Cola cool. With the other hand he takes potato chips, produced in Australia or perhaps further afield, from the plate on the table. And he tells me how good it would be if we could only monitor the consumption habits of indigenous rainforest peoples.

Introducing consumption

A review of studies of consumption reveals a diversity of approaches and definitions, straddling many disciplines. Fine and Leopold (1993: 3) present a succinct taxonomy of analytical perspectives on consumption:

> The discipline of economics ... argues that consumption bestows utility, in sociology, it may well be status or social position, in psychology, it is a conditioned response to gain a level of wellbeing, in anthropology, it has been interpreted in terms of its symbolic role in ritual ... More critically, consumption can also be viewed as a passive response to the goods that manufacturers offer, with tastes manipulated to guarantee sales and profitability.

There is a longstanding interest in consumption within anthropology (Douglas and Isherwood 1978; Appadurai 1986; Miller 1994, 1995a, 1995b, 1995c). Other works link consumption to a range of themes, including consumerism (Fine and Leopold 1993); political economy (Carrier and Heyman 1997); money, space and power (Corbridge et al. 1994); place, time and space (Urry 1995); technologies (Silverstone and Hirsch 1992); identity (Miller 1987, Friedman 1994); and nutrition and food intake (Hladik et al. 1993, Koppert et al. 1993, Pasquet et al. 1993). Bourdieu (1984) conceives of consumption as taste and choice enacted through social difference. In discussing the political economy of consumption patterns, Escobar presents the much-cited statistic that, '81 percent of energy consumption,' is by industrialised countries comprising only, '26 percent of the [world's] population' (1995: 212–13).

For the purpose of this chapter, consumption is defined as the amount of materials, goods and services humans use in living their lives and the symbolic meanings evoked by the quantities and entities used. This is placed in the context of a global environmentalist discourse of finite 'resources' and the impacts of industrial production. From both an environmentalist and an anthropological perspective, the study of consumption reveals important trends in terms of social injustice and environmental degradation.

Within this framework, and on the basis of contrasts which became evident to me during ethnographic fieldwork, consumption is considered in a comparative perspective here. I am concerned principally with comparing the activities of two groups: local people and conservation biologists. I refer to the autochthonous inhabitants of the Pio-Tura region, who are mostly self-sufficient, as local people. The scientists, researchers and assistants from outside the Pio-Tura area, who have been involved in the establishment of the 'Crater Mountain' conservation project, which subsumes a large part of the region, are referred to as conservation biologists.

It is not my intention to essentialise either of these 'groups'. They are each made up of individuals from different backgrounds who practise a range of activities to varying degrees. The ethnographic components of this chapter are intended to reveal something of the complexity of the human landscapes in which they interact. When juxtaposed, either theoretically or in actual situations, the two 'systems' of production, consumption and distribution which they represent are in stark contrast.

Many of the conservation biologists I interviewed and worked with did not appear to make linkages between over-consumption in wealthier, industrialised countries and global environmental destruction. At a personal level, they often failed to make conceptual linkages between consumerist actions in the lives of individuals, the systems of production within which they live, the differentials of wealth between donor and recipient countries, and the impacts of these factors on conservation. Within the context of so-called community-based conservation projects (see Western and Strum 1994), the absence of notions of complex interconnectedness at multiple levels can have far-reaching impacts, both on local people and on conservation itself.

Conservation and development: Consumption as power

The Convention on Biodiversity, signed at the United Nations Conference on Environment and Development (the 'Earth Summit') at Rio de Janeiro in 1992, linked conservation with 'sustainable use' (Sachs 1993: 3). It also defined 'indigenous knowledge' and 'traditional lifestyles' in terms of their use-value for conservation and development and encouraged national governments to do the same (Strathern 1999: 183). This and Local Agenda 21, the resulting action plan drawn up by over 170 nations with the aim of tackling global environmental and economic problems conjointly, took the union of conservation and development from the rhetoric of Rio to the experience of local people around the world. People, as well as biodiversity, had to be 'managed'. This reinforcement of development ideology in the context of the environment (Luke 1995: 75) has led to conflicting values and policies at the heart of conservation.[3]

Protected areas have often been based on Euro-American models of 'preserving' nature in the form of national parks. These were often propelled by a strategy of keeping people out, dating back to the colonial era (Mackenzie 1988). From the mid-1980s onwards, it became more 'fashionable' (Brandon et al. 1998: 1) to focus on ideas of sustainable use. It was no longer acceptable for local people to be ignored, hence the adoption of a language of participation in 'community-based conservation' of the 1990s (Western and Strum 1994). This was underpinned, however, by a seemingly irrevocable marriage of conservation and development: the 'Integrated Conservation and Development' (ICAD) project, 'introduced' in a World Bank publication in 1992 (Wells et al. 1992: ix). In this document, three defining 'operations' of such projects are outlined, in this order: 'Protected Area Management activities', 'buffer zones' and 'local social and economic development'. Management activities include 'biological resource inventories and monitoring'.

With regard to the language of conservation, Sachs (1993: xv) writes: 'In recent years a discourse on global ecology has developed that is largely devoid of any consideration of power relations, cultural authenticity and moral choice, instead, it rather promotes the aspirations of a rising ecocracy to manage nature and regulate people worldwide'.

ICAD projects have proved to be a complicated business. As Wells et al. acknowledge, 'Promoting local development is a highly complex and challenging task for conservation practitioners' (1992: x). Conservation biologists often engage in activities and undertake workloads which far exceed what they are trained to do (Takacs 1996: 2, Brandon 1998: 437–8).

In the context of wildlife conservation, humans are often presented as 'threats' to biodiversity. Although participation is often at the level of rhetoric rather than practice (Chambers and Richards 1995: xiii), at least the experimental moment in 'community-basedness' in conservation does not leave people totally out of the equation.

A further experimental development in conservation and environmental policy threatens, in some ways at least, to overshadow this, however. Carbon mitigation – otherwise known as carbon swapping, carbon offset, carbon sequestration, carbon sinks, emissions trading, trading permits and trading quotas – is a potential future for conservation in the tropical world which could eclipse local prerogatives. Stuart and Sekhran (1996: 1) refer to carbon mitigation as a 'novel type of economic opportunity [for developing countries]: as suppliers of a commoditised environmental service of green-house gas mitigation'.

Carbon swapping works on the premise that wealthier nations, or industrial concerns in those nations, give money to tropical countries with expanses of forest. In exchange for the forest remaining intact, they receive credits for producing greenhouse gases, believed to exacerbate human-induced climate change,[4] which they can produce in their own country. This circumvents the issue of the reduction of domestic consumption.

Luke (1995: 57) applies Foucauldian perspectives to the study of environmentalism. He refers to the capacity of discourses of nature and the environment to exert what he calls 'geo-power' over and through nature in contemporary economic and social governance. Even before nature is altered by technological change, it is transformed discursively through linguistic representations, in terms such as 'natural resources'. Once such discursive processes become normalised, Nature can be used, 'to legitimise almost anything,' (ibid.: 58). Environmentalism becomes an act of constructing and policing space, and the environment is reduced to sets of measurable components.

This is a form of power resulting from particular linguistic representations of 'environment'. The kinds of power which emanate from development-oriented forms of conservation also have procedural, as well as linguistic and discursive, origins. ICAD projects in places such as the Pio-Tura region are monitored vigorously. This gives practitioners of conservation and development the power to scrutinise the consumption practices of local people in the same way as they might wish to scrutinise behaviour among bird populations. Carbon swapping involves a further transfer of power. It enables rich nations to continue to expand their regimes of consumption and pursue unfettered economic growth whilst investing in commoditised tracts of rainforest[5] and exercising control over land and human behaviour in less wealthy nations. In this sense there is almost a direct correlation between consumption and power.

Consumption in the Pio-Tura region

When I am asked how people of the Pio-Tura region subsist, I find it difficult to give a sound-bite answer to encapsulate their human ecology. If I am asked whether they are hunter-gatherers, the answer is yes. If the question is whether they are swidden horticulturalists, this is certain. If I am asked what their staple food is, the answer is definitely sago, derived from the long-term cycle of planting, nurturing, felling and processing sago palms. Yet sweet potatoes and other tubers produced in gardens are also important foods. People also practice a number of other activities. They fish for eel, edible frogs and other freshwater creatures in small lakes and rivers. They practise pig husbandry. They sometimes keep chickens or cassowaries or marsupials from the forest in a domesticated setting. They harvest tree grubs, sago grubs and other insects from rotting timbers or sago palms. They harvest fruits and nuts from planted and tended groves in fallow gardens and from single trees in the forest. They collect green leaves, either from cultivated greens gardens, distinct from other gardens, or from areas or single trees in the forest. They harvest the eggs of birds and, at lower altitudes, turtle eggs, when in season. They share and exchange constantly with others to whom they are related or with whom they have important relationships, such as that formed by the pending exchange of women in marriage between two groups. They smoke meats from game caught on hunting trips – pig, cassowary, marsupial, flying fox, birds, snake, lizard. Women make string bags for carrying babies, firewood and garden produce, and sago bags for processing and carrying sago flour. Men make spears and bows and arrows. Timbers, barks and leaves are harvested from the forest to build houses of varying degrees of permanence depending on their function and the labour available. In short, it is a busy existence, characterised by the constant undertaking of work.

Anthropologists and human ecologists have warned of the potential to romanticise such people and practices, and a tendency to see the inhabitants of rainforests as being in harmony with nature (Ellen 1986, 1993; Milton 1996: 31). Slater (1995: 129) argues that such representations of Amazonia have practical consequences for the future of the people and place, in that they both 'dehumanise through idealisation,' and form the basis of misguided development programmes (see also Adams, Chatty and Novellino, this volume). According to Filer (1991: 23), such an idealised conception of the relationship between 'forest-dwelling' people and their 'natural environment' does not apply in Papua New Guinea. On the other hand, research in historical ecology has revealed intricate and inextricable links between people and their environment, indicating that humans are neither straightforwardly destructive towards the environment nor singularly biodiversity-promoting (Balée 1998a). The extent to which humans have a complex impact on the environment, and vice versa, and the degree to which nature is a 'human idea' (Cronon 1995: 20) are basic starting points here.

People in the Pio-Tura region seem in general to be eager to consume industrially produced food on the rare occasions when they have the opportunity. I recorded many comments, however, which expressed concern about both the nutritional and aesthetic worth of such foods. A woman in her forties in a family I was close to often said that store food does not fill you up. In 1999 I interviewed women at the market in Haia. When questioned on the good and bad sides of store foods, one woman in her twenties remarked 'Beer is the bad side. Men drink and get drunk and they abuse the whole family'. Another woman in her twenties said: 'Lollies and sweets rot children's teeth. Children see sweets and they want them and cry'. A common perception of town people who eat an urban diet was that they are fat (in a pejorative sense) and lacking in strength.

In discussion, people often said that they do not like a new food the first time they taste it, although it might become more palatable over time. Men who had worked for extended periods at a biological research station, where the staple diet was rice and tinned fish, complained about the lack of local foods. Their comments indicate that people miss their staple foods after only a short time, and this has significant effects on their feelings of well-being. Yet it can be said in general that people in the Pio-Tura region have predominantly positive associations with industrially produced food, especially if consumed on an infrequent basis.

Through the promotion of income-generating activities, one of the direct results of the conservation project on Pio-Tura lands has been the establishment of a second trade store, which sells a range of industrially produced 'staples': matches, rice, tinned fish, tinned meat, cooking oil, soap and so on. There has been an increase in consumption of such goods during the past few years. Wrappers and packaging are often discarded on the ground around the village, suggesting, to a Euro-American eye, the cultural specificities of the concept of 'litter'. On the other hand, packets and tins are used as toys or utensils by children or as decorations in people's houses.

There seems to be a common perception locally that the forest is not inexhaustible. Stocks of game especially are said to have become depleted. This is also remarked upon by biologists. It is difficult to assess with quantifiable evidence, other than repeated anecdotal accounts, that this is the case, and the reality is surely a complex one, which will need to be considered with care in the context of both conservation and development.

What I would wish to emphasise here is that people of the Pio-Tura region have a broad subsistence base derived from a huge range of activities. People are also eager to eat any type of food, either locally produced or introduced. Every part of an animal is eaten by someone – even the skin of pigs or the wings of bats. What is not eaten by people is passed on to dogs and pigs. Industrially produced items are passed between people in a similar way. They are divided and distributed, and even the wrappers and tins are used for something before they are discarded freely, perhaps as used materials from the forest are discarded.

This tendency to make use of all parts of something is reflected in methods of food production. Almost every part of the sago palm is used in its production, for example. Such a mode of subsistence, and indeed its social relations – the exchange, distribution and redistribution, which is a part of food production and consumption – stand in marked contrast to industrialised production, distribution and consumption. On a global scale of environmental degradation caused by over-consumption, the local means of procuring a livelihood from forest land appears to be more 'sustainable' than industrialised living, with the proviso that subsistence practices may or may not have detrimental effects when undertaken on different scales and in combination with other technologies. Another proviso is that the relative 'species richness' of each ecosystem is a significant issue from the perspective of global conservation. Yet we must be mindful of the power relations underlying a Euro-American assessment of indigenous modes of producing and consuming.

The Crater Mountain conservation project

In histories of the establishment of a conservation project in the Pio-Tura region, it is often emphasised that local people expressed concerns about depleting stocks of birds of paradise to expatriates working there in the 1970s (Pearl 1994: 199, Johnson 1997: 398). In brief, this process is reported to have led to the establishment of a Papua New Guinean NGO called the Research and Conservation Foundation (RCF). To varying degrees since its inauguration in 1986, it has been steered both in terms of ideology and financing by overseas donors, mainly from the United States.

The region which became known as the Crater Mountain Wildlife Management Area comprises lands of two distinct groups – Pawaia people of the Pio-Tura area, and those known as Gimi to the north. These two groups, inhabiting different expanses of the land demarcated for conservation, speak unrelated languages (Moseley and Asher 1994: Map 34) and have distinct histories, customs and modes of subsistence. Conjointly, parts of their lands, spanning a range of altitudes from highlands over 3,000 metres to lowland river systems about 100 metres above sea level, became recognised as a protected area within national law in Papua New Guinea in 1993 (Pearl 1994: 196, Johnson 1997: 397). Since the early 1990s, it has assumed the status of a high profile 'Integrated Conservation and Development' project, testing new methodologies in conservation.

Between 1995 and 1999, within the framework of the experimental movement of 'Integrated Conservation and Development', the project was further moulded by a programme entitled the Biodiversity Conservation Network (BCN). This was an initiative of the Biodiversity Support Program funded by the United States–Asia Environmental Partnership (US–AEP), a conglomeration of business and governmental interests led by USAID, the United States Agency for International Development. As such, it formed a part of the American input to the United Nations Global Environmental Facility. The US–AEP 'enhances environmental protection and promotes sustainable development in Asia and the Pacific by mobilising U.S. environmental technology, expertise and financial resources' (Biodiversity Support Program 1997: 120).

The local impacts of this programme in the Pio-Tura region have been wide-ranging. The Crater Mountain project was directed along a path to test a 'hypothesis' that raising the monetary income of local people through activities which might promote a conceptual link between biodiversity and cash could be beneficial to conservation (ibid.: iii). During the years of BCN funding, biologists and staff working on the project set out to 'increase the average annual per capita income of clans' (Johnson 1997: 397). They did this by promoting principally three activities: waged labour in support of biological research, the sale of local 'artefacts' for cash, and ecotourism.

Industrial consumption viewed from the Pio-Tura region

Since the 1980s, the Crater Mountain project has attracted a gradual stream of biologists to the Pio-Tura region, mainly engaged in research for masters and doctoral degree programmes in the United States, Australia and Europe. This has brought unique patterns of industrial consumption to a region considered to be one of the more remote parts of Papua New Guinea.

It is not the first time that local people have had contact with industrial ways of producing and consuming. A significant proportion of men left the area in the 1960s (Wagner 1979: 148) to work on coastal plantations and other commercial ventures of the colonial era. They saw towns and in some cases they were taken on tours of urban establishments to give them an idea of how an industrial system might work. Some men travelled south to Baimuru and the Gulf of Papua and saw other influences there. Other men have subsequently worked for logging or petroleum companies. A handful of women have had prolonged stays in towns, mainly due to the need for hospital care.

A missionary couple has been settled in Haia since 1973, and they have been bringing industrially produced goods into the area since then. My own research project has also depended to a certain extent on an industrial system of production and consumption. The difference between other external influences and the biological research, in terms of consumption and behaviour, has been in both scale and procedure. On the one hand, some of the biologists who worked in the Pio-Tura region in the 1990s showed sensitivity to local ways of producing and consuming. Some of the Papua New Guinean researchers and staff placed there have adapted to a local way of life, making only slight modifications according to their own needs or comfort. The practices of certain biologists have been at the other extreme, however.

Although each individual clearly behaves differently, a set of precedents has been adopted for the amount and type of industrially produced food which should be taken into the region for consumption. This has led to what might be termed 'package biology'.[6] Services and set procedures are offered to visiting scientists. As well as giving guidelines on how much food to import, this code of practice is designed to determine the working and economic relations visitors have with local people and also, indirectly, their social interactions.

Visiting researchers and ecotourists fly into the villages of the region by small aircraft or helicopter. They must bring with them sufficient cash to pay for the following: a fee for entry to the conservation area; 'sleep fees' (Johnson 1997: 415) in guest houses, research stations or local dwellings; the carrying of their provisions and equipment to the site of research by local men, women and children; and wages for the local people who act as their assistants, whether as guides, or in data collection such as finding birds' nests, measuring frogs and so on. They are advised to freight into the conservation area all provisions for themselves and for any people who are working with them whilst they are in the region, as it is deemed their responsibility, under conventions of the conservation project, to feed them. They buy these provisions from warehouses and supermarkets in the highlands town of Goroka. A typical shopping list, in bulk, includes: bags of rice, crates of tinned mackerel, tinned meat, noodles, cooking oil, sugar, chocolate drink, milk powder, breakfast cereals, oats, popcorn, pasta, tomato sauces, curry powder, salt, pepper, chilli powder, chocolate bars, sweets, soap, matches and bales of toilet paper. Although some of these goods are produced in Papua New Guinea, the vast majority are imported. A consignment of 're-supplied' goods might fill anything between a single Cessna aircraft (holding about seven passengers or 400 kilogrammes of freight) and a succession of Twin Otter planes (each holding twenty passengers or 1,600 kilogrammes of freight). On the occasion of an arrival of freight for a long-term research project, or one involving many people, such as localised biological inventories of flora and fauna, or for a staff meeting or annual general meeting to be held in a village, the amounts emerging from the aeroplanes are prolific, even to a Euro-American eye.

During the period of my research I encountered no expatriate biologists who incorporated local foods into their diet to a substantial degree while working there. The lack of local, as opposed to imported, food consumed had a significant impact on social interactions between local people and biologists. Through not eating local foods, such key aspects of local practice as sago production and consumption, for example, were not embraced by visiting biologists. In fact, both the sago palm and subsistence-related work were perceived and represented in a negative light, as disturbances to biodiversity (Johnson 1997: 396). The local market was frequented by local people, Papua New Guinean professionals (teachers, health workers, community development workers and biologists), the missionaries and myself, but not on a regular basis by expatriate biologists during the time I was there.

One visitor engaged in a biological research project told me of how she was primed on the food she should take with her by a biologist with longstanding work experience in the region. She was accompanied by the biologist on a shopping trip for bulk quantities of food to last for several weeks, and was advised on the amounts and brands of goods purchased. Kind though this was, the shopping ethic seemed to be based on 'treats', she said, and the result had amounted to little short of gluttony. Conversely, members of her team were delighted to be introduced to some local foods towards the end of their stay, and they wished they had been informed and encouraged to incorporate local foods into their diet during the period of research.

Every biological research expedition clearly has practical needs. Certainly, some provisions need to be shipped in, and the health and well-being of the researchers, who are often experiencing the extreme physical and psychological shock of finding themselves in an unfamiliar and relatively hostile environment, are of paramount importance for the success of the project. However, the precedents currently followed do not facilitate engagement with local ways of doing things. The 'package biology' approach ignores local history and subsistence and imports cash-oriented, consumerist modes of producing and interrelating.

This has subtle effects on how local people think and on how they aspire to consume. It also appears to have negative spin-offs for conservation. As outlined above, one result of raising people's cash income has been the establishment of a second trade store in Haia village. People have a greater appetite and purchasing power for industrially produced goods. It is often argued by project staff that this pre-existed the conservation project and that consumer desires are an inevitable result of a global economy. Yet instead of resulting merely from interaction between local people and a global economy, these desires have been promoted and accelerated by 'conservation and development' initiatives. Meanwhile, consumption is also promoted in the form of grants, international flights and careers for biologists.

In addition to the scale of change in consumption generated by the Crater Mountain project, there has been a marked impact on social relationships. A price tag has been attached to all activities and relationships and human labour, material culture and place have been commoditised. Behavioural precedents, including those which influence distribution of food (a key element of local sociality), have been imported as a cheap replica of consumption in the industrialised world. Expatriate biologists often experience difficulties in their relations with local people as both parties tend to become overly focussed on the goods imported rather than the relationships they might embody.

One day, two and a half years after I first arrived, the mother in the family I lived with said to me: 'Back home, you just get things from the store and eat things out of bags and tins, don't you?' We were talking about the difference between food there and in my country, and I was saying how much I would miss the local foods she gave me. Of course, there are elements of truth in her comment. Most foods consumed in my country have been purchased, most are packaged. For her, there was no imaginative possibility that the two forms of production and consumption could be similar. I tried to emphasise certain parallels: that I shop at a market with fresh produce, just as I do there, that I do not eat out of tins, and that we have our own staple foods. Her comment says a great deal about consumption in a comparative perspective.

Real connections

In the 1980s, deep in the forests between the highlands and the south coast of New Guinea, in territories previously uncharted by Western science, a site was chosen for a biological research station. Although, as oral narratives on its history would have it, the choice of its location was heavily influenced by the need to rest induced by a severe bout of malaria in an expatriate biologist, it has since become a symbol of nature and biodiversity untouched by humankind, of archetypal 'primary' rainforest.

In 1998 I visited the location myself. I was surprised to find a house which was palatial by local standards – the largest dwelling in the area at the time, with a spacious kitchen, living area, bedrooms and work area. It had a white porcelain flush toilet. About forty metres away were two basic dwellings for local people who came to assist biologists in their research. I thought of the houses built throughout the history of the region, and reflected that the scientists' house, with its tin roof and impressive size, must have had the greatest human-generated impact on the surrounding forests. Yet houses have come and gone for centuries there. They rot back into the ground when they come to the end of their natural cycle, and the tin roofing, like the timbers still intact, can always be re-used in the construction of another dwelling.

What tend not to be mentioned in terms of human impact on the ecology of the region are the pits of industrial waste from over a decade of biological research and accompanying meals made from imported food-stuffs. The biological research station in untouched primary forest is also a landfill site for consumer waste. So this is a human landscape.

I will not forget the conversation I had with a man whose mother was from that piece of land as we left the research station to go and sleep at his uncle's house, itself not a long walk away. When we were literally metres away from the metal-roofed house, he pointed out plants and trees which his mother's brother had planted or which he himself had planted and tended, and the spot where he had killed a wild pig. Clearly, in spite of the conception held by biologists that this is an environment untouched by human influence, this is an anthropogenic landscape *par excellence*.

It is ironic, then, that there are regular disputes between resident biologists and local people who fell ancestral sago palms on land which is set aside from human impact as part of the conservation 'deal'. This raises the question, 'What are the negative impacts of planting, tending or felling a sago palm?'

I once had the pleasure of felling a sago palm myself, in a human-use area within the conservation scheme of things. I worked with the family I lived with to fell and then to strip the palm. I was astounded by the many larvae, beetles, other insects and snakes nestling within its crevasses and in its soft bark. On no other occasion during the years I worked in those forests did I see so much entomological diversity gathered in one place. The felled sago palm is also a unique habitat for sago grubs,[7] a highly valued food. It is fixed in my memory as a harbour of biodiversity.

Before I ever went to Haia, I had the good fortune to meet a senior member of an American conservation organisation which was instrumental in the founding and development of the Papua New Guinean organisation RCF. We were in Port Moresby. I was on my way into the country and she was on her way somewhere else, having paid a brief visit to the conservation area sponsored by her organisation. She enthused about Haia and its surroundings. You only had to walk out of the village for two minutes, she said, and you were in primary forest. In an aside with a colleague, she mentioned the local custom of women making string bags from the fibres beneath the bark of the *tulip* tree. 'They should make a plantation of *tulip* trees,' she said.

I still reflect on the many questions raised by her statements. What, in her perception, made the forest two minutes' walk from the airstrip in the human settlement of Haia 'primary' and in some sense untouched by humans? What constitutes a primary forest environment, in a biological sense, and does it in any way acknowledge human activity? What makes the vision of a *tulip* plantation attractive for 'conservation and development' whereas the staple plant, the sago palm, is defined as a threat to biodiversity?[8]

Over two years later, I sought her out again, during a visit to New York. It was my intention to trace some of the linkages in terms of cash, ideas and personnel between people in the forest surrounding Haia and others scattered across the globe. We chatted for an hour and a half in a small diner in Manhattan. During the discussion she mentioned, with subdued excitement, the possibility of arranging a carbon mitigation for the Crater Mountain area, of finding a huge sum of money as a conservation pay-off. She said that there were many uncertainties and that these would have to be worked out before the idea could be put to the local people. She also intimated that the bureaucratic and financial process scared her. Although I reflected on the ethics of trying to arrange a transnational and theoretical transaction for a piece of customary land in Papua New Guinea, I recognised her dilemma. It might well be counter-productive to enter into lengthy explanations of the mechanics of carbon swapping before the ramifications of such a transaction are clear.

The ensuing moment, adapted from my fieldnotes, is reproduced here:

When she looks at her watch and makes moves to go in to work, I ask if I might accompany her, as I would like to see where her organisation is based. After two minutes' walk, we turn into an underground garage and she gives the parking attendant her car keys to deliver us a shiny red Mercedes, her customary mode of transportation. As we sit in traffic jams on the way to work, I ask about the car, interested to know what kind of linkages she makes between her own journey in stops and starts across New York City (not noticeably quicker than taking the subway) and her career in global conservation. She describes the car as a tool of necessity and explains something of her personal life and the decision to have and use the car. Like many people with an environmental conscience, she was faced with her husband's dilemma of his journey to work taking three times longer by public transport, so they decided to purchase and share the car. During the journey, we come to discuss people 12,000 miles away in the region of Papua New Guinea to which we both have a connection. In a rare moment of relative speed along a highway with several lanes during that traffic jam ride, she tells me of the undesirable effects of coffee gardens on biodiversity in the Crater Mountain area.9

Conclusion: Paradoxes of environmental practice

I have drawn attention to a number of ironies revealed through a comparative study of consumption practices of local people and conservation biologists in the Pio-Tura region of Papua New Guinea. In conclusion, I propose to reconsider these ironies and to explore their significance for local people and conservation.

The model of consumption promoted by 'package biology' is a parody of expatriate subsistence. Flying around in helicopters and aeroplanes with large quantities of store goods is the image of expatriate life presented to local people. At the same time, incoming conservation practitioners monitor local behaviour. In doing this, they contest customary rights (to harvest sago palms, for example) and tend to depict local people as threats to biodiversity. Equally, local people are encouraged to adopt the stereotyped expatriate model of consumption. When working with biologists, they are fed on tinned and processed foods. 'Income-generating activities' are promoted through waged labour in a Western image, albeit in the form of relatively menial tasks paid below levels of Western minimum wages. Meanwhile, local custom and landscape are commoditised for research, ecotourism and the sale of artefacts. Visitors are encouraged to consume culture and place. In terms of social justice, carbon swapping heralds a further transfer of the costs of resource use. Rainforest and its inhabitants become consumables in attempts by wealthier nations to avoid having to make politically unpopular choices to cut domestic consumption. This has implications for ownership and control of resources where high levels of biodiversity are found.

Thus, a picture develops of conflicting rights to consume (see Nygren, this volume). The conflict between customary land tenure and 'conservation and development' is acted out in disputes over landscapes and their components. Conservation is propelled by a sense of moral urgency, both to save species from destruction and to 'develop' local people in order to reduce their impact on localised biodiversity.

While consumption practices based on a Euro-American model have been promoted in a locality where even basic health and education services are lacking,[10] there are many alternative paths to development. A pragmatic approach to achieving conservation and sustainable development might aim at finding ways of meeting the articulated needs of local people within a joint framework of environmentalism and social equity. Initiatives based on health and education, on local terms, might be more appropriate to subsistence livelihoods than the promotion of ICAD. With a modicum of interdisciplinary research and planning, such initiatives could be established to meet reasonable local desires for development.

Where does the right to consume have its origins and in what terms might we think about it? This chapter suggests that empirical ethnographic study can throw light on some of the ironies of this question. Ethnography facilitates comparison between apparently separate spheres, helping us to explore links between policy and practice, to consider the impacts of policy on local people, and to recognise the paradoxical tenor of some of these interactions. Focussing on consumption is a way of highlighting inequalities of power in perceived rights to resources and imbalances of scale in their use. From an ethnographic standpoint, it might seem that the quest for solutions in which perspectives of local people are prioritised (and perceived from a local viewpoint) is more of a responsibility than a right.

The juxtaposition of descriptive passages, discussions and overviews of conservation history and policy mirrors the process of ethnographic revelation concerning the complexity of social and environmental practice. Underlying assumptions become clearer. People and nature are often considered to occupy separate spheres, for example. An ethnographic lens shows that they clearly impact on each other, and that those promulgating the view that people and nature should be kept separate evidently make their own impacts on the nature they are reifying.

So the ethnographer, through a series of discussions and interactions, is often introduced to the complexities of environmental issues in striking, paradoxical, even absurd ways. There is scope for bringing these interactions to the fore, for evoking them and integrating or contrasting them with other levels of analysis (see Brosius 1999). This also highlights the many contradictions of being an ethnographer in an age of eco-speak, or the environmental sound-bite (see Cronon 1995: 22). On the one hand, the demand for polarity in political debates tends to evade the inherent complexity of issues concerning people and conservation. On the other hand, many ethnographers are only too familiar with the struggle to depict complexity.

The point of the comparisons made here is not to critique or reward the consumption practices of particular individuals and produce a 'sustainability ranking' for consumption levels. We cannot classify individuals or groups as being environmentally friendly or destructive to the environment according to a homogeneous measure. The point here is rather to provide a commentary on the complexities inherent in global conservation and the relations of production, consumption and distribution from which it has arisen. I am also concerned with how conservation operates in relation to local people in Papua New Guinea. These data suggest that global conservation and local practices have become interlinked in terms of consumption and the inequalities between them. There are many unresolved paradoxes in environmental management in which we are all implicated. In spite of the difficulties they pose to ethnographers, engagements with these questions might benefit both local people and conservation. Although conducting ethnography of processes such as those described here can be a sensitive undertaking, this chapter suggests that this is necessary if anthropology is to be allowed to apply its findings to the achievement of more equitable environmentalisms.

Acknowledgements

I would like to thank the people of Haia and the Pio-Tura region, pilots and staff of Mission Aviation Fellowship and New Tribes Mission, Dr Colin Filer, Michael Laki and colleagues at the National Research Institute, and staff of the Research and Conservation Foundation and the Wildlife Conservation Society. Without assistance from and collaboration with these people in Papua New Guinea, the research which forms the basis of this chapter would not have been possible. I am grateful for comments on earlier drafts from David Anderson, Eeva Berglund, Louise Dawson, Roy Ellen, Christin Kocher Schmid, Nicholas Thurston, Paige West and three anonymous reviewers. I would like to thank the other participants of the workshop at Goldsmiths College in April 2000 for stimulating and inspiring exchanges. I acknowledge invaluable funding from the European Commission DGVIII through the Future of Rainforest Peoples Programme (*L'Avenir des Peuples des Forêts Tropicales* – APFT).

Notes

1   Given the sensitivity of the topic of consumption and the form of comparison adopted, I have employed a number of techniques to conceal personal identities whilst maintaining attention to narrative detail. I wish to emphasise a range of issues and connections between them and not the identities of particular individuals.

2   This might be seen historically as a conflict over the 'resources' of the Panguna copper
     mine. The mine, and the environmental degradations associated with it, became the cen-
     trepiece in a struggle for the island's independence from the state of Papua New Guinea
     (Filer 1990).

3   Critics of the concept of sustainable use in conservation (Ehrenfeld 1988, Dove 1993,
     Sanderson and Redford 1997, Crook and Clapp 1998) maintain that market-oriented
     approaches perpetuate the very mechanisms which have adverse effects on biodiversity.

4   Mabey et al. (1997) consider the political economy of climate change and international
     economics, including what they call 'tradable permits/ quotas' in carbon dioxide (ibid.:
     30–1). The idea of 'marketable pollution permits' is not a recent phenomenon. Most prac-
     tical experience of trading of pollution permits has been between States in the United
     States (Pearce and Turner 1990: 117–9). The Kyoto Protocol, the result of the Convention
     on Climate Change held in Japan in December 1997, is a proposed code for industrialised
     nations to reduce their emissions of carbon dioxide and five other greenhouse gases by 5.2
     percent below their 1990 levels between 2008 and 2012. Proceedings from a subsequent
     conference (Institute of Petroleum 1998) discuss the Kyoto agreement and carbon trading
     from a business point of view.

5   A language of free market economics has been employed in debates about carbon swap-
     ping, most notably by resource economists. McCallum and Sekhran (1997: 45) refer to the
     rainforests of Papua New Guinea as a 'bank' of carbon-fixing carbon dioxide, coherent
     with the conception of both people and forest as providers of a 'service' for which there is
     a market.

6   Research and Conservation Foundation/ Wildlife Conservation Society (1995: 49) refer to
     'research service packages' to be offered to visiting scientists and students.

7   Tuzin (1992: 104) refers to the sago grub as *Rhyncosphorus ferringinlus Papuanus*. He also
     documents some of the 'biodiversity' to be found in or on the sago palm, and the many
     human uses of the palm for Ilahita Arapesh people in East Sepik Province of Papua New
     Guinea.

8   People of the Pio-Tura region do plant and tend *tulip* (*Gnetum gnemon*) trees in fallow gar-
     dens, along with a variety of other palms and trees. The research I conducted with Pawaia
     people on their human ecology suggested that there is a concept which is perhaps the local
     equivalent of 'plantations', although it represents something closer to forest permaculture
     than monoculture. One of the words for this is *teijano*, or 'food area'. The leaves of the *Gne-
     tum gnemon* tree and other parts and fruits of a range of planted and tended trees and
     palms are important foods in the local diet.

9   The sale of coffee beans is one of the ways for people in rural areas at mountainous alti-
     tudes in Papua New Guinea to generate their own cash income to meet health and
     education needs.

10   See Research and Conservation Foundation (1999) for an example of how this was
     recently acknowledged by project practitioners.

PART III

WRITING ENVIRONMENTALISM

8

'WE STILL ARE SOVIET PEOPLE'
Youth Ecological Culture in the Republic of Tatarstan and the
Legacy of the Soviet Union

Luna Rolle

Introduction

Resistance to the Soviet state to a great degree is articulated through the
space of Russian environmentalism. As Gupta and Ferguson (1997:19)
argue, any resistance to power is linked to processes of cooptation and complicity. Given the prominent role that environmental movements have
played in the dissolution of the former Soviet Union, Russian environmentalism is simultaneously conservative and presses for change. This chapter
deals with the structure and ethnography of a specific ecological movement
in Tatarstan, a Turkic republic in the southern part of European Russia. The
data for this chapter came from six weeks of fieldwork in the city of Kazan
where I worked with an ecological student group called the Guard of
Nature. The research consists of twelve translated and transcribed interviews, and many informal discussions, particularly with my Russian–English
interpreter. The quotations from activists, analysed within their historical
context, helps to understand power and underprivilege.

Following the end of the Cold War, can Russian ecological culture now
be understood within the single framework of global environmentalism? Or,
do Russian ecologists hold a unique view of the environment, differing both
from Western capitalist approaches to 'nature' and alternative views coming
from developing countries (Guha and Martinez-Allier, 1997)? In order to
answer these questions, I will explore the way in which both culture and the
notion of global relationships are contextualised within the geopolitical categories bequeathed by the Cold War. First, I will argue that post-Soviet

environmentalism is similar to Euro-American environmentalism in the way it uses the notion of 'global'. Second, I will establish that the Guard of Nature uses the concept of 'nature' in order to frame political arguments opposed to the status quo in a manner also typical of Western social movements. Finally, and most importantly, I will use material from my interviews to show that despite these common links the discourse of the Guardians can only be understood within the unique social-political legacy of the Soviet Union and of the Cold War. Thus, I will show that while the legacy of the Cold War gave the Guardians a certain 'Western' view of global relationships, Russian ecological culture in Tatarstan grows out of a unique regional history of ecological underprivilege.

The socio-political context of environmentalism has been studied extensively by Milton (1996). According to her, Western-centred theories of environmentalism are based on the economic opposition of North versus South, elaborated by the North and imposed on to the South. She stresses the idea that environmentalism is a Western invention. Countering her view, Guha and Martinez-Allier (1997) argue that environmentalism in the South is an expression of subordinated social groups, originating in social conflict over access to, and control of, resources. In this chapter I will establish that environmental discourse in Tatarstan reflects both processes. As in many Soviet republics, Moscow planned to build several nuclear stations in the territory of Tatarstan. After the Chernobyl catastrophe, a local ecological movement developed against the project. This single movement had a crucial role in the awakening of environmental consciousness generally in Tatarstan. It was under conditions of regional underprivilege that an ecological movement became concerned with other major environmental issues in the Soviet Union. Yet, despite the local background to the movement, the interview data clearly show that the Guardians rely upon globalising discourse in order to express their ideas and their opposition.

The Guard of Nature in context

Under the Communist regime, ideological obstacles impeded the discussion of the environment. The 'road towards Communism' implied industrialisation. Thus, environmental degradation was in a way evidence of the construction of socialism. Despite the evidence of very serious environmental problems, environmental issues were only given recognition by state authorities at the very end of the Soviet period, and, even then, they were not allowed to become politicised. It must be stated that Russian society was not totally indifferent to ecological problems before the reforms. The Guardians recognise their predecessor in the Volunteer Environment Protection Squad (*brigada*) [VEPS], a student movement which arose in the 1960s at Moscow State University. It was also in the 1960s that the global environmental crisis was revealed to the public as a concern of workers in capitalist countries. The VEPS was composed of intellectuals studying ecology and was practically unknown to the general public. However, this protest group can be considered as the first sign of an awakening consciousness of the ecological situation in the Soviet Union. It provoked the formation of a similar movement in Tatarstan, as Galina Ivanova, member of the Ecological Network of Tatarstan, recalls:

> The student movement has a long history. Thirty years ago, a Department of Environmental Protection was created at Kazan State University. The Student Squad for the Protection of Nature was then founded on the wave of other student Squads, mainly the Moscovite one.

In the same period, some independent currents emerged within official state organisations. The Guard of Nature was founded as an independent organisation in 1972 but had to gain legal recognition in order to be recognised as a social actor, as explained by Sergei, leader and founder of the Guard:

> The first seminar of the Guard was held in 1972 ... In the middle of the 1960s, some environmental information which had never been published before started to appear in the press. There was very little of it, but by collecting from different sources, reports, newspapers articles, etc., it was possible to create a picture of what was happening. Censorship was very severe and we could not publish our first article. Then, analysing the situation, we understood that in order to say something publicly, we needed to enter the legal and 'official' frame.

The aims of these organisations were to save the environment and, at the same time, to promote democracy. There was a strong connection between the ideas of ecology and democracy in state socialist systems (Yanizkii 1987). In a regime which did not recognise legally independent organisations, the fact that a group of students revealed the inability of the state to solve ecological problems was a challenge to the party's authority. Sergei explains how the Guard could at the same time avoid censorship and act effectively:

> From the beginning, our task was to show that the governmental machine as a whole was useless and unable to solve ecological problems. How did we do that? We chose the information which was in different publications and we compressed it into short reports. These reports showed that the ecological situation was catastrophic and worsening. Moreover, many arguments were distorted by our press. For example there was an article in the official newspaper on 'Making the city green', comparing the situation of Tatarstan with the West. There were some data about the situation abroad, but no statistics about our situation. There was a sentence saying 'Aren't the 15 square metres of green per one inhabitant in Vienna too small?' we printed the sentence and under it wrote our own 'In Kazan, there are 7.8 square metres per inhabitant, according to the report of the local leader, Comarad Lipatnikov' with no comments. Nobody could contradict it.

In the early 1980s, the volume of industrial production in the Soviet Union reached its peak. Soon, the serious effects of environmental degradation started to profoundly affect the population and could no longer be hidden. The overuse of chemicals resulted in mushrooming allergies and illnesses among children. The dacha boom took place: urban dwellers acquired a second home outside the city in order to spend their vacation and weekends 'in the open air and close to nature' (Pickvance et al. 1997: 213). The policy of glasnost began in 1985 and permitted activists, to a certain extent, to receive the necessary formerly 'secret' information about ecological matters. In 1986, the Chernobyl disaster marked a turning point for the growth of mass ecological movements. As Iskander Yesaveev, professor of sociology at Kazan State University recalls:

> When Chernobyl happened in April 1986, we did not have enough information, because glasnost was a very slow process ... For example, in May 1986 my family and I had decided to go on vacation to the Black Sea. The river next to Chernobyl flows in the Black Sea and we postponed the vacation to the following August, because we thought that by that time the water would be clean enough. So, there was information about it, but the seriousness, the scale of the disaster was not really understandable ... Later, we understood that the scale of the catastrophe was huge, and that the consequences of it could be very serious for Russia, Ukraine and, of course, for Tatarstan.

By the end of the 1980s Russian ecological movements achieved a high level of mobilisation. Environmental problems such as the drying up of the Aral Sea and the pollution of Lake Baikal by paper mills were revealed in the Soviet press. The 1986 campaign against the Siberian rivers diversion and the 1987 campaign for the protection of Lake Baikal illustrate the generalisation of mass ecological awareness in the Soviet Union. The number of groups, associations and movements with a larger focus than green issues increased dramatically and expanded into 'political clubs' (Leningrad) and 'social initiatives' (Moscow). In 1987 the Brateevo and Bitsa Green movements emerged in Moscow, constituting the first mass ecological movements in Russia (Pererjolkin and Figatner 1997). Local movements in Moscow and St. Petersburg then developed into 'popular fronts' in both cities by the late 1980s (Yanizkii, 1987). The following year, these 'fronts' applied for and obtained official status and registration for local and national elections.

From 1991 onwards, with the development of general political parties and the appearance of conflicts within ecological groups, the membership and the level of political activity of ecological groups declined and their focus returned to specific ecological issues. More importantly, as explained by Anderson, the situation of economic collapse following the end of the Soviet Union affected all social classes, making personal economic survival the first priority. The development of the Guard of Nature generally followed these stages. Its activities were radical until the early 1990s and particularly during the *perestroika* period, when the movement could freely express itself. It fought against poachers and participated in ecological marches based on the model of the internationally known Greenpeace.

After the collapse of the Soviet Union, its activities became closer to those of an association for nature preservation than those of a radical ecological movement: it stopped getting involved in radical actions, and coming into direct conflict with the institutions. As Sergei explained:

> [Since the end of the Soviet Union], we have not been in direct conflict with official institutions. One of the main reasons is because we have never done actions similar to those of the Rainbow Keepers. Although we did take part in protest actions, our task always aimed at involving in the guarding of nature as many different people and organisations as we could.

The movement today organises environmental education through clean-up campaigns and summer camps. It sends volunteers to speak to students in schools. Every year they collect rubbish on the banks of the Kazanka River and try to make the public aware of the problem of pollution.

In 1999 the Guard was not a mass movement but a local student organisation relying on a regional framework of similar organisations. The Guardians themselves are generally hostile to suggestions for direct and illegal action. The methods of action adopted have become 'civilised'.

Legal social action ('using dialogue' with authorities) is preferred to mass protest:

> The State did everything right about ecology but it executed things in the wrong way. Sometimes there is a lack of democracy in this sense. I am trying to act legally because nobody wants to risk too much, but if there are problems, they have to be solved, sometimes, according to my own motivation (Jena).

> The Guard took part in illegal actions but not me. This action was conducted by Rainbow Keepers, this movement is famous for its cruel methods, concerning both the law and the participants of the action. They go against the law and nothing can stop them. I think they are right when they argue against the law, but we do the same not with physical methods but by using dialogue. After each action a small tragedy occurs, like broken hands or legs ... I do not think this is loving Nature (Gulnora).

Problematising the global view

The Guardians belong to a web of world-wide exchanges of ideas and policies. They have successfully appropriated, and transformed, global environmental discourse. However, it has to be noted that this model of environmental discourse has been generated by Western institutions operating at the global level. The former Soviet Union was only partially included in certain environmental debates such as the UN Conference on the Human Environment in Stockholm, 1972. Therefore we should not be surprised to find significant differences in the way the Guardians relate to and incorporate the sometimes contradictory issues of 'development' and of the 'environment', as illustrated by Denis' discourse:

Humans should have a close relationship with natural environments, as they used to have in ancient times, 5,000 years ago. The depletion of natural resources is the result of development. In the 1950s and 1960s, we were told that within twenty to thirty years everything would be done by machines. But where are we led to? So, I agree with the scientific revolution but its achievements should not be used in such a negative way.

On the one hand, Denis' last answer exemplifies two globalist assumptions: that development (in the sense of 'technology') is positive and that it has to be made sustainable, i.e., not used 'in such a negative way'. On the other hand, there seems to be neither a devaluation of 'non-developed' societies nor the assumption that 'being developed' is the only way to be. Rather, the image of a 'primitive' society, like '5,000 years ago', relating to nature in a 'better' way serves as a model. We will return to the Guardians' appeal to the 'primitive myth' later in this section.

The globalist view, according to Milton (1996: 188), usually implies several 'anti-globalist' assumptions, such as the idea of opting out of the global economy and of placing the management of resources in local hands. These assumptions are often linked to a suspicion that 'development' favours wealthier sections of the population.

The assumptions of the 'anti-globalist' approach are only partly shared by the Guardians and this is explainable by the fact that their limited ecological knowledge (in ideological terms) does not permit elaboration of a framework which stands in opposition to the globalist model. Many, consistent with the quote from Denis above, participate in the assumptions of the dominant model communicated through the media. Some, like the chief 'ideologist' of the movement, Sergei, articulate a more complex view which is closer to the 'anti-globalist' message that one would expect:

> To solve the global ecological crises, we need more than a radical change because the existing economic and political mechanisms reflect people's orientations towards a certain type of life styles. We need to change both tastes and people's personal priorities. Since the system is based on long historical traditions and subsequent events it is very difficult to change it and we can only do it in a certain frame, at a particular time. This evolution is possible and now we are analysing different tendencies existing in several spheres of society and probably in a year or a year and a half we will come up with a theory accumulating all these elements. Towards year 2000, the net of communities has already started to appear in our country, like the type of community named *longama* and other communities which existed in the West.

Both the globalist and the anti-globalist perspectives aim at conserving the environment 'as a resource for human use'. The third perspective, radical environmentalism, defined as 'ecologism' by Dobson (1990: 13) and 'ecocentrism' by Eckersley (1992), is based on the understanding that the environment has a value independent of its use to human beings. This view requires us to see ourselves as just one element in an ecological system rather than as the centre of the universe. It is possible to identify elements of this view in the discourse of the Guardians. It is implied in what Sergei would call 'the crash of the contemporary civilisation':

> Sometimes I think ['a crash of civilisation'] would occur if everybody's mentality would become like that of our Guardians. In that case the world would become horrible from ordinary people's perspective [laughter]. Everything would be destroyed, all modern culture, pop art, the entertainment industry, the car indus-try, the modern economic system would collapse. The radical change of the consuming system would result in the crash of contemporary civilisation.

What is unique in the way that Sergei combines ecologism with a critique of globalism however, is the necessity of a break between the past and the future. Sergei understands such a radical change as signifying a break with the Soviet past and a rejection of a consumerist 'Western' future *before* it has been built.

In line with ecocentric precipts, all the young Guardians share the view that 'nature' has an intrinsic value, and some of them place it in a superior position:

> Nature has a value in itself and this value is bigger than the value of all human lives. I think that humans took the power, without asking, of considering them-selves superior to all other species, but who knows what will be in a million years? (Denis).

> Nature has a value in itself because nature has its own laws and these rules are not subordinated to humans. Men always tried to put nature in a frame but they could not (Natasha).

Despite these classic ecocentric ideas, the Guardians combine their view of nature with a faith in a myth of 'primitive ecological wisdom'.[1] It is this element which distinguishes the Guardians most strongly from Western ecological movements. The Guardians often praise the ability of non-industrial people to live in harmony with their environment. They see this as somehow 'natural' or 'innate', as if these communities possess more ecological wisdom than those which, through economic development, have become alienated from 'nature'. They therefore identify the industrial economy, as the root cause of environ-mental problems. The nostalgic view of a primitive past in which human beings lived in harmony with 'nature' appeals to the Guardians:

> I think people should go out of the city, into 'nature'. People cannot live in houses made of bricks all their life long. People need to go out in 'nature' and rest in it. Wooden houses are better than ones made out of bricks (Natasha).

> People did not exploit 'nature' as much as they do now, they consumed 'nature' but their possibilities were limited. With the scientific revolution in the last sev-enty years, it became very obvious that 'nature' was being spoiled and it affects my soul. The further we go with scientific progress, the more we ruin ourselves (Jena).

It should now be clear that the Guardians articulate elements of global environmentalism but that they combine these with a view of 'nature' can be considered as part of the Western 'Nature conservation perspective' (Milton 1996:124–5).

Personal and collective change

In this section I will clarify why the Guardians' culture can be defined as 'oppositional'. The Guard, to a certain extent, can be characterised by its opposition to the Soviet past. However it is important to note that the Guardians negate the Soviet past in a very deep and fundamental way in a manner which makes their outlook very similar to that of environmental activists in the West. Like Western movements (Ashford and Halman 1994) the oppositional outlook of the Guardians is highly individualistic in contrast to the collectivism of the Soviet period.

It is ironic that decades of Communist rule have given rise to a modern, secular and Westernised generation that is neither supportive of Communist institutions nor wedded to tradition. To understand this, it is important to remark that the Guardians belong to a generation which did not experience the October Revolution, collectivisation, the Stalinist purges, the 'Great Patriotic War', post-war reconstruction and the Soviet-led crushing of the Prague Spring in August 1968. The young Guardians, born in the 1970s, were moulded by more recent events: the war in Afghanistan (1979–89), Gorbachev's reforms and the Russian intervention in Chechnia. In the last decade of the regime, the West became a reference point by which young Russians judged their own society (Dobson, 1994: 233). Following the 1986 Chernobyl disaster, growing awareness of the magnitude of environmental pollution and the system's inability to cope with it alienated the Russian youth from official youth Communist groups. Freed by glasnost and *perestroika*, environmental groups gained the support of thousands of young 'greens' around the country. Surveys show that young Russians have gone further than their elders in repudiating key elements of Communism, such as the monopoly of the Party, egalitarianism, and opposition to private property. They also value freedom of the press and political pluralism (Dobson 1994:248).

More importantly, the Guardians belong to a generation that is experiencing the combined impacts of environmental ruin and cultural uncertainty. These factors make the Guard an oppositional movement that lacks deep ideological roots. One can say that it is the condition of environmental underprivilege which leads the Guardians to imagine a cultural politics of saving nature.

The individual motives for participation place the Guardians' culture within the broader oppositional culture of Western environmentalists. Socialising is an important motive for participation in environmental groups. In the Russian context, where youth could not be involved in activities outside the 'formal' sphere until a decade ago, socialising carries a high value in the discourse of the Guardians. All of them join the Guard partly 'to make friends':

First it was a place to socialise but now I got a lot of knowledge from the move-ment and a possibility to use it (Denis).

At first when I entered the Guard, we were good friends and, except when work-ing, we had a good time together (Gulnora).

Acting in the Guard gave me some information of course, but it is by talking to people and getting to know them that I acquired knowledge in terms of social relations (Jena).

In social terms, I found some friends here (Natasha).

The Guard shares a second fundamental characteristic with Western social movements. Its members challenge the existing order of things by nurturing alternative lifestyles. The term 'lifestyle' involves deep moral and existential questions of 'self-identity' – issues of how people should live in emancipated social circumstances, as described by Sergei:

The Guardians' point of view, similar to the one of Western radical ecologists – who are maybe even more numerous than here – is not compatible with the global mechanisms of contemporary civilisation. Our ecological ideal is living in an agricultural community, without the use of non-renewable sources of energy, supported by self-production and production for long-term use.

What is interesting in this statement is the process of 'dematerialising' the economy. Rather than struggling over material resources, the Guardians are engaged in a struggle over culture and meaning (Morris-Suzuki 2000: 65). In the post-Soviet context, they strive to deliver a new cultural message about the search for a better future. The action today is justified by the hope of a better world in which 'nature' will be taken care of, as explained by the activist Natasha:

Some people like Sergei Germanovich just cannot live without guarding nature. He always says, 'How will you live in the future if you do not look after Nature now?' This is the way we live in the Guard (Natasha).

Another motive for individual participation is the general belief in the usefulness of one's actions and a deep personal commitment. The Guardians present themselves as the bearers of a 'cultural' message about environmental issues: the concern for 'nature' is in their 'personality' or in their 'soul'. Nature helps people make connections between their own little lives and greater things:

My concern for environmental issues comes from my personality. I am always going against and if everybody says 'It's black', I will say 'It is white'. And even though everybody thinks the ecological situation is fine, I will say that it is impos-sible that everything is fine (Gulnora).

Ecology is in my soul. I think that if ever I was born for something, it is for Ecol-ogy. If a problem exists someone must solve it (Jena).

Even though there is an acknowledgement of self-interested reasons for wanting to stop something they regard as locally noxious to them, like water and air pollution,[2] it can be argued that the creation of an ecological environment 'for oneself' automatically means 'for all'. The spirit of collectivism and solidarity is indeed essential. Joining efforts and a collective approach to environmental problems are factors consolidating initiative groups from inside. For the experience of resistance to alter the identity of subjects it has to be connectable to some form of practice and representation (Gupta and Ferguson 1997: 20). Dr Yesaveev participated in the 1989 environmental protest against the building of a nuclear power station in Tatarstan and explains here what the reasons were for his involvement:

> I think the reasons I got involved were not 'globally' oriented, it was indeed a very concrete movement. First of all, we were supporting very practical reasons against the construction of this plant ... It was planned in a very dangerous and seismic place. After Chernobyl, we understood that it was very dangerous because we knew about the low level of technological control and equipment.

Yesaveev's experience illustrates a characteristic of all social movements: the 'multiversality', i.e., the ability and the will of the Guardians to articulate at the same time specific local experiences and forms of knowledge which address common global concerns (Morris-Suzuki 2000: 73).

A conservative, yet dynamic movement

In this last section, I would like to generalise from the experience of the Guardians to examine the special nature of Russian ecological culture. Here, I am most concerned to identify the continuing influence of contextual elements in its dynamics.

A common, indivisible identity

The Guardians generally acknowledge their common involvement in ecological issues and express a sense of belonging to the same 'community'. The apparently immediate experience of community is in fact constituted by a wider set of social and spatial relations. To be part of a community is to be positioned as a particular kind of subject, 'similar to others within the community in some crucial respects and different from those who are excluded from it' (Gupta and Ferguson 1997: 17). Most members acknowledge differences in the way they connect to ecological issues. These differences, however, unite rather than divide them. Belonging to the Guard does not involve assertions of primordial identities of blood, community or religion, but rather 'invented communities' with which people choose to identify themselves in a temporary and provisional way (Morris-Suzuki 2000: 68). The shared knowledge and experience are necessary to create a collective story that encompasses individual differences and nurtures a common identity.

There are obviously social differences within the membership but the Guardians themselves do not consider that this division implies any inequality. Indeed, the concept of 'social stratification' is very much inherited from the Soviet Union's official stratification:[3]

> Approximately, the social origins of the members is similar. There are children of military officers, intelligentsia, engineers, and, as far as I know, there is nobody with peasant origins. There are children of working class too here. However, there are no social divisions in the Guard, we are all a big, friendly family (Denis).

> The members are mainly students and people who graduated from our university. The social strata is homogeneous (Sergei).

A Soviet 'technocratic' approach to 'nature'

It has been shown how the Guardians contribute to global environmental discourse in the way they connect to natural environments. However, the Soviet conception of 'nature' helps to explain why the Western Green rejection of the pursuit of infinite economic growth is not what Russian ecologists are mainly concerned about. The Bolshevik view of natural environments found its roots in an intellectual current – the *Westerners* – which developed in the late nineteenth century (Ziegler 1987: 7). Their approach to 'nature' was a utilitarian and 'technocratic' one. The new order that they envisioned was modelled on Western Europe and led by the vanguard of society – scientists, industrial workers and engineers. Science and reason were the answers to Russia's social problems, along with an overall confidence in humankind's ability to control and transform the natural environment.

By rejecting peasant populism in favour of 'scientific socialism', Russian Marxism fell solidly into the technocratic tradition. Marxism's materialist perspective considers economic growth in an historically determinist way, as inherently progressive: technological development is the causal agent of social and political development:

> An agrarian society in which nature was not radically remoulded to suit civilisation's needs was by definition inferior to an industrial society ... Economic and material prosperity were by definition good (Ziegler 1987: 8).

Human beings, before becoming producers, were seen as living in harmony with their environment, like wild animals. The capitalist system, based on private ownership, alienated human beings from each other and from 'nature'. Communism was imagined as the state of perfect homeostasis between humans and nature. It could only be attained once humans overcame the alienation of the product of their own labour (capital) and therefore eliminated their alienation from 'nature'. This ideological framework had major practical implications in the Soviet context: it assumed that

Soviet socialism, by achieving 'social mastery' over 'nature' through technology, could cope with environmental degradation without challenging its major economic and social assumptions.

It can be shown that the Guardians share some features characteristic of the official Soviet image of the environment. They acknowledge the fact that natural resources are finite, but their faith in science and technology brings them to a rather ambiguous conclusion: the scientific and technical revolution that poses unprecedented environmental dangers also provides the potential for solving problems:

> The biggest problem of our Republic is the pollution of the ground water in the South-East. In this district dozens of villages have not had a drop of drinkable water ... Everybody there is carrying three litres of water in glass vases; people are going far away in other villages to collect drinkable water from pure springs. People borrow water from each other. This picture is apocalyptic. It is a very complicated problem but it can also be solved in technical terms. From the technical point of view, you just need money to solve the problem (Vadim Marfin, Professor of Natural Sciences at Kazan State University).

As a result, resolving conflict over environmental matters should, for the Guardians, ideally be based on technical considerations. This faith is very much a heritage of Soviet official propaganda that stressed the merits of socialism. The direct consequence of this first assumption is that the Guardians do not generally identify technology and economic growth as environmental threats *per se*. Indeed, the ecological crisis is seen as something that can be solved within the framework of a market economy and through profit:

> There is a need to change the system, or maybe the ecological crises will be solved if there will be someone who will find it profitable not to have this rubbish or to create national parks. If this person will say something intelligent, then everybody will get interested and find it profitable (Natasha).

> I think that the market economy is right, but I consider that it should not affect the ecological balance. To solve the global ecological crisis everything must be done according to the law (Jena).

Individualism as 'de-responsibilisation' – or abdication of responsibility

In the previous section of this chapter, I showed that young Russians accept the emergence of individualism as somehow natural. However, the growth of individualism in Russia does not seem to correspond with an increased sense of individual responsibility and an 'ethic of commitment'. The Guardians' discourse does not escape to the collectivist mentality inculcated by more than seventy years of Communist propaganda. The interviewees shared a general tendency to expect politicians or experts to solve environmental crises. A survey by Koliesnik and others showed, for example, that both experts and the general public identify the 'irresponsibility of economists' as the main cause of environmental problems, after 'a shortage of money' (1995: appendix 2).

As mentioned above, Russian youth supports political pluralism and private property. However, democratic institutions are not firmly rooted in the post-Soviet context and private enterprise skills have been non-existent. A powerful example of this process is the Kafkaesque way bureaucracy worked under the regime and frequently functions now. Sergei's account is illuminating in this respect. Under the Soviet regime, when Sergei and his friends decided to undertake environmental action, they needed to talk to the Trade Union Committee (*profkom*) supervising all the university associations. As Sergei knew that the *profkom* would find an excuse not to help them, the most direct way to obtain results was to write directly to the City Executive Committee (*ispolkom*), supervising the *profkom*. The *ispolkom* was glad to find people who would work for free on environmental matters and wrote the *profkom* 'an angry letter' urging them to help Sergei and his friends.

Conclusion

This exploration has shown that: (1) the Guardians incorporate and transform the notion of global; (2) Russian ecologists only partly challenge the status quo, but that they are creative in the way they invent, transform and reinterpret images of 'nature'; and, (3) despite these common themes, Russian ecological culture does not fit in any of the Western-centred North-versus-South classifications. Let us consider the three conclusions in detail.

The inclusion of the Guardians' culture in global environmentalism is crucial for understanding its dynamics. The Guardians are social actors who simultaneously experience the local and the global. Therefore, rather than opposing an autonomous local culture to a homogeneous movement of cultural globalisation, I tried to show the way in which the Guardians pick up and transform common cultural forms. The political construction of 'nature' and the appeal to the 'primitive myth' elaborated by the Guardians place them within a broader global environmental discourse. To analyse the Guardians as an oppositional movement then, is to place them more generally within the frame of Western environmental movements. However, at the same time it is important to underline the group's innovative character, and to recall how previously any political self-activity was supervised by the state and official organisations linked to the Party.

My main hypothesis was that the elaboration of environmentalism in the Guard of Nature inherits Soviet values and norms. I argued that environmentalism does not challenge the dominant set of values inherited from the Soviet Union, but rather incorporates them and, through this process, serves the maintenance of the social order. Its young members function within a framework which is still embedded in Soviet values and this legacy, I argued, greatly influences their ways of seeing environmental issues and their responsibilities towards 'nature'. For example, the dichotomy between 'nature' and 'culture' was at the base of the Soviet ethic of industrialism, which aimed to scientifically master 'nature' for the well-being of the masses. This legacy deeply influences the Guardians' views, in their faith in technology and 'scientific progress'. They react to the bureaucratisation, the rigidity and the incompetence of authorities, but never oppose the latter politically. Their primary aim is to advise and reform 'official' organisations. Ideologically, they reject the monopoly of the Party but not the desirability of continued economic growth.[4] Therefore, the 'alternative paradigm' elaborated by Russian ecologists does not challenge but rather reinforces the dominant social paradigm.

However, I wish to reiterate that in the post-Soviet context, the struggle of the Guardians as environmentalists takes place over the realm of meaning and identity rather than over material resources. As in Europe, it becomes increasingly 'cultural' in character (Morris-Suzuki 2000: 67). As in Nygren's contribution, this ethnography is about competing interests in the contested representation of nature and culture. It is important to remember that the Guardians belong to an urban movement that is not dependent on natural resources. For them, nature is idealised and disconnected from the social needs of the activists. Therefore, it is the richness of metaphors of 'nature' created by the Guardians that needs to be emphasised.

Environmentalism in post-socialist Russia shares many features with environmentalisms elsewhere, but models derived from the Western world fail to do justice to its uniqueness. As a highly industrialised country, Russia, like the former Soviet Union, is embedded in the technocratic Marxist model, which aimed at managing nature for human use through technology or 'scientific socialism'. The idea that nature is a resource for human use is common to both Russia and capitalist countries. In this sense, Russian environmentalism shares very little with environmentalist views held by populations in the South (Guha and Martinez-Allier 1997). Indeed, in contrast to the local people in the Pio-Tura region of Papua New Guinea (Ellis this volume) for instance, these urban Guardians have incorporated a highly industrial culture and do not suffer from a colonial past (unlike, arguably, rural Tatars). The Guardians enjoy a certain freedom of action given that there are no outsider non-governmental organisations (NGOs) attempting to represent them. They are also in a sense privileged by their residence in an urban regional capital.

This chapter has argued that it is important to remember that the Russian variant of environmentalism is not the same as that of the Western world. There is a significant gap between social-science models of environmentalism and the environmental issues arising in the former Soviet bloc.

Post-Soviet environmental logic cannot be understood within the terms of 'global' environmentalism, or even North-versus-South models. The Guardians are caught between the distressing realities in their immediate surroundings and the ideological 'world' that tells them to think globally. More importantly, their experience of ecological underprivilege leads them to consider what possible futures are open to them in a very rich imaginative way.

Notes

1   The debate over the primitive myth has been exemplified by Diamond (1987) and Johannes (1987). It has to be noted here that the demonstration that the myth of primitive ecological wisdom is not true is now well-founded: it is known that there are non-industrial societies that do not recognise a human responsibility to protect the environment. Other factors reducing the impact of people on their environment, such as geographical isolation, low population density, and limited technology, have been stressed (Ellen 1986). Finally, it is clear that industrialism *per se* is not the cause of environmental destruction: cases of lavish wealth and conspicuous consumption are known in non-industrial societies. See, for example, the Trobriand islanders (Malinowski 1935) and the Kwakiutl of Vancouver islands (Howard and Mageo 1996).

2   According to Koliesnik et al. (1995 appendix 2), acute water pollution and air pollution are believed to be the most important problems in Tatarstan.

3   The Soviet Union officially divided the population into three social classes: the peasants, the workers and the intelligentsia. The status of each of them depended on their political closeness to the Communist Party of the Soviet Union. The upper class was, in real terms, the *nomenklatura*. The intelligentsia and the employees were the middle class, the workers were considered the 'working' class, and political prisoners and criminals were at the bottom of the hierarchy.

4   The absence of claim to limit the production of material goods, except by the leaders of the Guard who know about and are obviously influenced by Western environmentalism, can be explained by the fact that such a view can hardly be popular in Russia. With the experience of economic dislocation and a dramatic drop in living standards since the collapse of the Soviet Union, environmental concerns are generally relegated to a secondary position.

9

THE ECOLOGY OF MARKETS IN CENTRAL SIBERIA

David G. Anderson

The tempestuous 'transition' in the former Soviet Union has threatened the livelihoods of people from all walks of life (Bridger and Pine 1998). Since the end of Yeltsin's presidency, those who once crowned the status ladder, such as teachers, intellectuals and pensioners, now struggle to survive on starvation wages. Workers and civil servants in state-owned enterprises may wait months or years for back-pay in devalued currency. As the country fights a determined and increasingly desperate civil war in the Caucasus, military units in peripheral districts barter their rations and weapons for food and indifferently dismantle their barracks to heat themselves. Standing in contrast to the ranks of the newly dispossessed are those entrepreneurs who, having placed themselves at the front of the queue for redistributed state property, often live lives of wealth and leisure unimaginable to most of their fellow countrymen.

It should not come as a surprise to report that rural minority peoples in Siberia, not always at the centre of attention of the redistributive state, are also suffering distress from the collapse of the socialist economy. However, one would expect the indigenous fishers, hunters and herders of central Siberia to wield certain survival skills which could tide them over during the collapse of an industrial economy. Still others would expect, perhaps, that an end to industrial production, with its negative environmental consequences, would be one of the best things to happen to this remote part of the globe. In fact, it is this primitivist image of sober self-reliance which has made Siberian rural peoples some of the most underprivileged in the Russian Federation today.

The purpose of this chapter is to describe the ecology of market reform in Siberia. Specifically, I will question the relevance of the topographic metaphor of 'the market' which suggests that it is sufficient to map out a stock exchange or a village hall in order that food and money be distributed rationally. Instead I demonstrate that markets chart out networks of relationships in a manner more similar to the trophic chains and predator traps of modern ecological theory. Although this theoretical point has been made by other authors (cf. Polanyi 1957, Hann 1998), the ethnographic approach will reveal unexpected practical dimensions to this idea when applied to the social and economic crisis in eastern Siberia. Here, radical shifts in the economic relationships between people have already induced a new terrestrial ecology while attempts to define and protect an abstractly defined environment are creating roadblocks to the development of robust local economies. Here, as in the case of Bedouins or Bataks (Chatty and Novellino, this volume), parks and property transfers conducted in the name of democracy and conservation have so upset local ecological understandings as to create an unsettled environment which is nevertheless profitable for a few. These investments in economic change reveal a menacing side to the term 'sustainable development' (Escobar 1996).

The chapter is based on two visits to the Evenki Autonomous District (Krasnoiarsk Territory) in central Siberia as part of a two-year project supported by the Circumpolar Liaison of the Canadian Department of Indian Affairs and Northern Development and by the Russian State Committee of the North (Anderson 1998; Campbell 2001). The goal of the project was to transfer knowledge from Canada's three northern and predominately aboriginal territories about the marketing of 'wild' products, the formation of community development corporations, and the cooperative co-management of migratory wild caribou/reindeer herds. I have combined the immediate project objectives with my scholarly research interest on the meaning of privatisation and the so-called 'wild market' in Russia (Anderson 2000b). Together with research assistants Craig Campbell, Lise Wilson, and the director of the Department of Rural Economy, Artur Ivanovich Gaiulskii, we collected a rich set of data on the recent history of reindeer herding and the hunting of migratory caribou, market surveys on the local pricing of meat, and have surveyed the local technology for processing meat. These data were originally collected in order to advise international development agencies on how to design their programmes more effectively. However, they also inform anthropology as they quite clearly show the complex ecology of relationships which arise when people try to adapt to enforced programmes of privatisation. Before moving on to the analysis of how these market relationships integrate themselves in historical and social settings, it is important to situate Evenkis and the Evenki district geographically and historically.

The Evenki Autonomous District is a large rural district at the edge of the Putoran plateau of north-central Siberia and the wide taiga tributaries of the Yenisei River (see Figure 2). The District is a 'homeland' of sorts to Evenki, Ket and 'Yessei' Yakut peoples - all of whom at the turn of the century built a successful economy of hunting wild game, trapping furbearers, and using harnessed reindeer for transport. The road travelled by these people from the hungry days of the Civil War preceding the Bolshevik revolution, to the equally hungry days heralding the onslaught of democracy and capitalism, has been a long one. In the past seventy years the Evenki District has been industrialised by various mineral expeditions. Further, native peoples have been urbanised into small yet highly dense settlements which are dependent upon the supply of fuel and imported food. Most importantly, Evenkis and Yakuts have learned to manage the multifarious employment positions provided by Soviet socialism. The impact of massive changes experienced by four generations has been ambivalent, and some might say negative. As is well known, during the 1930s dozens of people in the region were executed for being wealthy herders (*kulaki*) or for deceiving the masses with their shamanic songs. Forced into residential education, children were de-skilled in subsistence techniques and knowledge of their own language. Yet the experience also made them into some of the most highly trained representatives of indigenous peoples in the circumpolar north. State-sponsored industrial reindeer husbandry disrupted the subtle ecological relationship between wild and domestic reindeer in the district, but it also provided several generations of Evenkis and Yakuts with a good living effected through time spent on the land living with reindeer.

Reindeer ranching was the keystone to the ornate artifice of state socialism in Siberia. In Party posters and texts, socialist reindeer herding was held to be one of the great gifts that the Soviet people offered the world and especially their indigenous cousins living in Alaska, Canada and Scandinavia who reputedly languished on reservations and on social welfare cheques. Giving substance to the slogans, practically all elements of the rural economy circulated around the reindeer. The fortunes of villages rose and fell by the accounts of how many reindeer were born and how many were slaughtered. The fame and careers of individuals were made and broken by heroic accounts of success (and failure) in reindeer ranching. At the end of each fiscal year, the accounts of collective and state farms were balanced by highly subsidised coefficients based around the value of reindeer meat. Literally, wages were paid, homes were heated and electric lights burned thanks to wealth on the hoof.

The sudden declaration of the bankruptcy of central planning and collectivised agriculture sent the Evenki district into crisis. Since the agricultural reforms of 1991, which made rural enterprises 'financially independent', the numbers of reindeer remaining in domestication have dropped to a fraction of their total during the Soviet period. More troublesome are stunning accounts of rural poverty, of hunters and their families not receiving wages or even pensions for months and even years. In Evenkiia for the first time in history there are now cases of rickets, not to mention tuberculosis, both ailments related to the lack of fresh meat and the lack of basic tools (ammunition, petrol) to hunt for food. Ironically, the direct cause of this crisis seems to have been reforms which were aimed at giving local producers complete independence from central government agencies and a propertied stake in their own herds and pastures. Thus this leads us to the question of whether radically unreigned market spaces now threaten what were once autonomous local economies of hunting, reindeer herding and fishing.

The short results of our research are fortunately that no, the current market transition is not necessarily a death sentence on rural economy in Siberia. Evenkiia, like most regions of Siberia, has many highly skilled and committed hunters, herders (or, as they are now called, farmers), and local processors of wild products, all of whom intuitively understand the obstacles but also the opportunities created by the political shifts of the past decade. We have also discovered that understanding this situation is doubly difficult for Canadians, or Euro-Americans more generally, quite simply because the common-sense definitions of 'the market' are very different in the Russian North than they are in the Canadian North, and are very different now within Russia than they were ten years ago. Nevertheless, and this is the important result of our work, we have found that there is a solid base of skill and curiosity upon which to build very useful conversations as to the relative value of tools and techniques, new community institutions, and the value of reindeer and caribou for human identity. In short, what the Russians call the 'wild market' has not destroyed or made irrelevant all which is descended from the Soviet period, but has opened up some new ways of integrating local and global interests.

It is important to qualify this statement with a disclaimer. Although this chapter presents a lot of detail about profit margins, supply and demand, and consumer willingness 'to pay' for reindeer 'commodities', the goal of this discussion is not to open the floodgates to commodity capitalism even wider than they have been already. The tragedy of the current so-called 'market' reforms in Russia, as with free-trade agreements across the world, is that they benefit a relatively small group of highly organised entrepreneurs and have led most working and rural-based populations into poverty. Nonetheless, trade liberalisation does allow local, face-to-face communities to try to set their own values on local products, like reindeer and caribou meat but perhaps not as efficiently as with the support of a sympathetic state structure. Thus our rule-of-thumb has been that if our analysis could help increase the volume of trade in reindeer and caribou products threefold (ignoring for the time that producers might be recompensed for only one third or one quarter of the value that their meat commands in urban markets) then local aboriginal people would live much better than if they had no income whatsoever. Further, since working with caribou and reindeer is a locally evocative undertaking, supporting this activity allows for younger generations to gain skills in being on the land and using the local language. Of course, a far *better* situation would be if these activities were supported directly, however this seems to be a rather idealistic goal given the extent of the degradation of life in the former Soviet Union.

The terrestrial ecology of post-socialist Siberia

In contrast to what one would expect, one of the first major changes brought by a centrally imposed policy of market reform has been an inversion of the terrestrial ecology itself. Today, the places of wild and tame *Rangifer* have become reversed such that most of those who are considered herders today are in fact hunters of wild reindeer. This should not be viewed as a serious contradiction. In central Siberia there has for many centuries been a continuous relationship between domesticated reindeer herding and the hunting of wild caribou. At various points in history, the numbers of wild caribou have been dwarfed by the numbers of domesticated reindeer. At this juncture in history, partly due to the deregulation of forests and people, the relationship is now the reverse.

Unlike in the European North, or in the western parts of Siberia, local herders have never had a well-developed tradition of keeping massive herds of domesticated reindeer for the production of meat. Instead, they kept reindeer to meet other needs such as for transport, for hunting wild caribou, and for ritual purposes. From the 1960s, the Soviet state invested great resources in trying to instil a system of large-scale reindeer ranching in central Siberia. While the official statistics show state planners were successful at getting the numbers of domestic reindeer to double in the 1930s, quadruple in the 1950s, and then sky-rocket in the 1970s and 1980s (Amosov 1998: 68–78), for the most part this was an unsustainable tendency for many intertwined social and ecological reasons (Anderson 2000a). In this heavily forested, escarpment-ridden area of central Siberia the maintenance of domestic reindeer herds of over 1,000 head, sometimes approaching as many as 3,000, required a massive infusion of resources to maintain helicopter supply chains, organised aerial culling of wolf populations, complicated 'shift-method' labour rosters, and scores of kilometres of fencing. These financial resources, while still present in Russia, are now being directed to purposes other than 'the socio-economic development of the Northern Frontier'. The collapse of the resources upon which Soviet reindeer ranching depended led to an explosion in the number of 'wild' reindeer and a sharp reduction in the number of domestic reindeer. While in Evenkiia there were approximately 60,000 head of domesticated reindeer in the 1980s and a modest population of migratory caribou of approximately 200,000 head, by 1998 there were less than 5,000 head of domestic reindeer (Gaiulskii 1999) and wild reindeer approaching, very approximately, one half million.

The precise reasons for this shift in what I am arguing is a historical relationship between wild caribou and domesticated reindeer, are not clear. Popular hearsay accounts sadly comment that local people simply 'ate' several thousand head of deer due to a lack of money to buy food. An equally prominent explanation, and a more reasonable one, highlights the effect of an escalating wolf population since the end of the centrally organised helicopter-based culling of the population. The third commonly heard reason, which is also reasonable, is that many thousands of deer became feral due to the fact that their human wards were busy queuing for social assistance payments in regional centres. With some sympathy for local herders, but entirely no evidence, I would hypothesise that reindeer herders may have deliberately culled or released many thousand of reindeer which were 'difficult' to keep in favour of maintaining smaller, more well-disciplined transport herds.

It is also important to remember that the outlines of the present 'crisis' remind the historically minded reader of similar inversions in the relationship between wild and tame deer in the late nineteenth century (Vasil'ev 1908) and during the Russian Civil War (Rychkov 1922). Indeed, older Evenki hunters and herders in both the Evenki Autonomous District and the Taimyr Autonomous District readily admit that times are bad, and perhaps as bad as they were during the Second World War. However, conditions of life even during this dramatic transition are still thought to be a bit better than the mass human and reindeer epidemics, as well as human starvation, at the start of the last century.

The main point of analysing this dramatic collapse in the numbers of domestic reindeer is not to argue that reindeer husbandry is in its twilight in central Siberia, or that native peoples are too lazy or spoiled to keep reindeer, as many Russian urban observers argue. Rather, I would like to emphasise that local Yakuts and Evenkis have shifted their skills to maintaining a minimum number of domestic reindeer and to discovering ways of exploiting a larger population of migratory 'wild' deer. The feral reindeer that they hunt are just in a less intensive relationship with people. If the 'wild market' could provide hunters with fuel, ammunition and transportation to urban centres one could say that local hunters might live rather well off this first 'privatised' wild resource. However, in addition to changing the terrestrial ecology of the region, so-called 'free' market reforms have also brought tight structures which control distribution and, most curiously, a conservationist ideology designed to protect wild places. To understand these effects on local lives, we have to examine the ecology of post-socialist markets.

The ecology of post-socialist markets

In what ways are migratory 'wild' or 'feral' deer still the mainstay of the local economy? In today's very difficult rural conditions, migratory caribou from the Taimura and Taimyr populations provide one of the main sources of protein for most villages throughout central northern Siberia (the other staples being moose and fish). In Evenkiia, roughly between 3,000 and 4,000 head of caribou are officially hunted and distributed or accounted for through administrative channels. No doubt another 1,000 to 2,000 are hunted for local use. To the north, in the Taimyr Autonomous District, where the city of Noril'sk provides a wealthy urban market with a better transportation infrastructure, the numbers of migratory caribou culled annually are about double at 8,000 to 10,000 head. According to the unanimous opinion of economists and biologists, there is no reason why the population of caribou could not supply double or triple the amount of wild meat. It is at this point that other infrastructural, institutional and technological bottlenecks, discussed below, begin to present themselves as problems.

Post-socialist conditions have not only inverted the relation of people to deer but they have also created a unique atmosphere of intrigue within which the trade in protein is situated. The existence of an unrealised supply of *Rangifer* venison in a region where there is a high demand for meat, has to be understood within the context of another relation which is the product of what Russians colloquially call the 'wild' market or 'wild capitalism'. Despite the fact that there is a large supply of readily available local fish and meat, the staple food of people living in urban centres, such as the cities of Krasnoiarsk, Dudinka, Igarka and Noril'sk, has become imported canned and fresh frozen meat (Figure 2). For example, our market survey for the capital of the Evenki District, the village of Tura, for two quarters of 1999 showed that at any time one could find in any one of ten stores in Tura (of which there are reputedly sixty) meat products from Moscow, Minusinsk, Krasnodar, as well as Korea, China, Belgium, Germany, New Zealand, United States and Brazil (Wilson 1999). The local explanation of this paradox is that these imported products, which were smoked, canned or vacuum-packed, or fresh-frozen, looked better and stayed fresh longer over the difficult transportation routes to the store. Perhaps more important but even more paradoxical was that these foreign products were cheaper than local produce by almost half. Another aspect, which is changing rapidly, was the perception that foreign products were of better quality than local produce.

A second major quality of 'wild' capitalism in Russia is that very little of the surplus earned goes into funding local processing and production. This is in dramatic contrast to the way that industries were constructed in the Soviet period when there was often a strong emphasis, almost an imperative, to develop local processing industries. It is important to draw attention to the *speed* of the collapse of the secondary processing industries. This implies that in most regional centres there is often a large supply of underused equipment and unemployed but highly trained personnel. This is a major difference as compared to regions like northern Canada. It also implies, for better or for worse, that most regional consumers are well familiar with the value of venison as a low-cost, low-quality staple, unlike in most parts of

Europe, the United States and Canada where reindeer or caribou is a high-priced 'exotic' meat. Thus when studying the problem of trying to re-commodify a sector, like the hunting of wild reindeer, there is a large social foundation upon which one can build, unlike in almost any other part of the circumpolar North.

When one adds both of these elements together, there is an interesting quality to the post-Soviet market in wild products: there is a potentially large regional demand for reindeer and caribou meat in most small centres and urban areas which at present is being met by relatively expensive foreign imports. Thus it would seem that there are some technical obstacles to the development of regional markets of local products which would in turn help provide cash to aboriginal households. However, is this really the case?

Although there has been a dramatic collapse of large-scale reindeer ranching and organised collective hunts of migratory caribou, these activities have certainly not ceased. Even in the desperate conditions of the late 1990s there was a steady, if clumsy process of supply of reindeer and caribou products to both regional and urban markets. Almost none of this trade is for money, but often for both notional and literal barter of goods and services. Nevertheless, the trade in reindeer and caribou meat is constant enough that one can identify certain regional regularities which can be used to describe the local ecology of economic relationships.

The most difficult place to find evidence of local trade in meat is in retail stores. Nevertheless, reindeer and caribou products appear irregularly on store shelves, depending on the season. In Evenkiia, for example, the primary season is during the spring (March to early May) when the migratory Taimyr population of wild caribou collects into large aggregations and makes a 1,000 kilometre journey north to tundras surrounding the shores of the Arctic Ocean. This is a liminal season for any slight change of the prevailing weather patterns might, within a matter of a few days, make ice-roads swell and melt frozen caches of carcasses stored throughout the taiga. Delivery prices, therefore, tend to be at their lowest as the producers live and work in a climate of risk. In 1999 the Tura Aviation Company made a great profit by buying meat from desperate 'farmers' at the fire-sale price of 8 rubles per kilogram and then realising the meat at retail prices of 25 rubles per kilogram in their own store, and some rumour as high as 35 or 40 rubles per kilogram on the Krasnoiarsk market. Prices for meat tend to rise through the summer as meat becomes rarer. Before the autumn hunt begins, for example, the lowest priced caribou product is the 'soup-bone assortment' at 20–25 rubles per kilogram and the most exclusive product is the boneless caribou roast at 32.50 rubles per kilogram (Wilson 1999).

To understand the meaning of these prices it should be noted that the average wage of a Russian worker in the relatively affluent setting of a northern village could not be expected to reach 5,000 rubles per month. Reindeer products competed against other imported proteins, most significantly pork, at prices substantially higher than that asked for reindeer. Thus pork sold for 45 to 60 rubles per kilogram (with bones) or 65 to 82 rubles per kilogram (unboned) while pork hocks and ground pork sold for 40 rubles per kilogram and 65 rubles per kilogram respectively. Fresh beef was not a common commodity, but steamed canned Chinese or Russian beef was considered to be a staple at 21 to 30 rubles per 500-gram can. Chicken, the next most popular food, would sell anywhere from 38 to 108 rubles per kilogram depending on the cut and country of origin.

The most expensive type of meat was semi-prepared or processed meat. This type of smoked or seasoned product, judging from Canadian experience, is also a natural market niche for caribou producers. The most popular product of this type was sausage (*kolbassa*) which was sold in a bewildering range of categories (some seven main types) from 60 rubles per kilogram for regional 'steamed' variants to imported 'smoked' variants at 85 to 90 rubles per kilogram. Similarly, cured pig fat (*salo*), as difficult as it was to find, sold at a similar price of 90.80 rubles per kilogram (Wilson 1999). It should be obvious that the advantage from a retailing point of view of the more expensive cuts is the fact that they are popular, pricey *and* they have a long shelf life.

This short-list of public, retail prices captures only a mere fraction of the actually existing caribou transactions and only a sliver of the volume of caribou traded in the region, but it suggests some interesting patterns. As one elderly economist told me in the Lands Administration office of the Ilimpei County Administration, despite any improvements to quality or taste, consumers expect that caribou meat will always be priced at half the price of pork, which according to my research and that of Lise Wilson, is an amazingly stable figure – having at least a ten-year time horizon. (Compare this figure to the rising relation between prices for vodka, bread and snowmobile parts in Taimyr [Ziker 1998]). This 'traditional' price creates difficult challenges for caribou producers since they must work against a price ceiling for a product which in turn depends on the much more dramatic changes in the price of transportation technology. I have not been able to find much reason behind this stable price relationship other than many hearsay comments to the effect that caribou meat is considered to be a 'working-class', 'unappetising', 'poor-quality' meat. Upon further questioning it became clear that these comments were linked to the association of caribou meat with institutional settings (state farms, cafeterias, hospitals, boarding schools, even prisons) which were not known for their quality control. If I made clear that I was interested in the meat itself, rather than the 'industry' connected to it, many Russians commented that the meat was 'lean (i.e., not tasty)', 'wild-tasting' or 'a natural product'. The last comment is the most interesting as it turns out that most post-Soviet consumers describe wild meat as a 'free' good which bears no human costs of being fed or otherwise cultivated. Thus it was seen as immoral that it was sold at all. It seems ironic that those qualities which make caribou venison unappealing or cheap to Russians are exactly what makes it attractive to high-priced urban markets in the United States and Japan. Recently, following the BSE scandal and reports that beef from the United Kingdom was surreptitiously imported into Russia there has been talk of caribou meat being 'ecologically clean'. However even more recent fears of heavy metal or radioactive poisoning have seemed to moderate this belief.

By far the largest sector of caribou trading in both Evenkiia and Taimyr is the quasi-commodified barter sector. The main parameters of this trade are the exchange of several 'standardised' stretched caribou carcasses (*tusha*) for barrels of fuel, spare parts and ammunition, groceries (including vodka), or often for transport. Rather more abstract examples involve the trade of meat to resolve chains of unpaid payroll debts. Most of these barter transactions, with the exception of the last type, are not registered. Thus it is very difficult to give an idea of scale other than the fact that 'everyone' recognises that this is how meat moves. Rough estimates are that the state agricultural sector of state farms harvests 1,000 head, private farmers harvest 2,000 head, and the oil workers in Baikit obtain a further 1,000 head.

Meat is bartered in mid-winter (January), when the overland winter roads open, and again during the intense spring period (April). An entrepreneur takes an all-terrain six-wheel-drive vehicle loaded with goods on a two-day journey of some four hundred kilometres to the places where caribou are being slaughtered. The vehicle is limited in its capacity for trade goods since it can only carry six to seven tonnes and must carry enough fuel to go there and back. Roughly half of its payload is fuel. Similarly, there is a physical limit to how many 'standard' carcasses it can load since the heads and legs sticking out in every direction limit its volume. A load of 126 carcasses seems average (five tonnes). The terms of trade are rather strict. In 1999 a two-hundred-litre barrel (*bochka*) of low-grade fuel would be traded for five carcasses. If the average carcass was forty kilograms in weight, this works out to the very simple figure of one litre of fuel per kilogram of meat! In the spring of 1999, fuel could be purchased for 3 r/L. In these terms of trade, hunters were 'earning' the lowest rate in the caribou market. However, much more unfavourable terms of trade can be had for one half-litre bottle of 'white fuel' for three carcasses.

Second to the trade in 'fuels' is the trade in transport services. It is said that one can negotiate transport of one hundred carcasses on a 'back-haul' trip of an all-terrain vehicle from Yessei to Tura for the price of twenty carcasses. A local trip to the hunting grounds in a caterpillar-track vehicle could cost three carcasses. Helicopter pilots purchase caribou on-site at the lowest price of 8r/kg.

Much more complicated chains of barter occur within state and semi-state enterprises. For example, the Department of Rural Economy in the spring of 1999 gave fuel (at 6r/L) and ammunition (at 4r/bullet) 'as an advance' to local hunters in the village of Yessei. As in the pre-Revolutionary days of the fur trade, the Department then collected this value at the price of 15r/kg. If extra caribou were presented they would be reimbursed in fuel or ammunition but very rarely in cash. In another arrangement, farmers were 'hired' in 1999 by the local administration to hunt caribou in lieu of unpaid back wages to local village teachers or other civil servants. The hunters would be reimbursed at the notional price of 15r/kg, most likely in the provision of transport services, or for fuel. A similar arrangement has been set up by the Tura Aviation Company which obtains caribou meat 'officially 'at a price of 10 r/kg and uses it to eliminate their unpaid salary rosters for 20r/kg.

Although crude in terms of quality and efficiency by Canadian or Scandinavian standards, this sector is an interesting proof that there is a supply–demand relationship and that it is met through very complicated means. Presumably, if a credit system existed, or if transportation technology was more efficient, the volume of trade and hence local employment of hunters could be increased many fold. However, what is most striking about the extent of actually existing barter relationships is that they show how reciprocity lies nested within the control of specific regional conglomerates, each of which tries to feed or sate its own workers in a paternalistic manner. Thus, to look at the regional caribou 'market' in another way, the fact that caribou protein is exchanged at all in the absence of credit, a reliable transport system, and a large disposable income in the hands of consumers is in fact a monument to the strength of social relationships in the region. In this example it is not necessary to peel off layers of symbolic fetishism to reveal relationships of reciprocity. The ecology of market relationships here, as in Russia in general, stands relatively unadorned with money values, margins or interest. Like exchanges of energy in a biotic community, the exchange of produce literally rests on the exchange of fuel in the form of protein, petrol or alcohol.

The ecology of market relationships, while wonderfully personalised, also favours certain predators who feed off specific communities. The regional trading groups, which created nested markets, are also interested in limiting their scale in order to be better able to control the way that commodities flow (but not necessarily to increase their value). If the market metaphor is to have any territorial meaning at all here, it is certainly not in terms of creating an environment where buyers and sellers mingle, but in defining the bounds within which exchange occurs. Thus, as the state has abdicated control of reciprocity and exchange we see the development of a fragmented market ecology where commodities do not so much flow as are rationed. Most agents active in the trade of meat are happy if they are able to feed their own workers with a supply of fresh, reasonably priced meat which they obtain at a low cost to themselves using existing inventories of machines. They have little interest in expanding their production to supply 'outsiders' and, more often than not, absolutely no interest in providing employment to aboriginal peoples generally (aside from those who are 'their own' workers). The only interest that people have in breaking out of their own networks is with the promise of what I identified as an extraordinary profit margin of up to 300 percent (Anderson 1998).

Thus, for example, the Department of Rural Economy has a large regional network of salaried 'state' hunters, a fleet of vehicles, a supply of fuel, influential contracts with the Tura Aviation Enterprise for freight transport, a new and hardly used meat plant stocked with imported Swedish technology, and access to central subsidies for the transport of inputs from the south. Although they have access to all of these resources inherited from the Soviet period, their total 'production' is the shipment of some 1,000 head of caribou (forty tonnes) on 'free' back-haul freight runs, the majority of which is 'sold' in a raw, unprocessed form to their workers and other civil servants. A small portion of this is processed in the Swedish meat plant to produce 200 kg of smoked sausage or roasts in a year (on a production line which can produce 100 kg per day!) The products are in turn 'sold' to their own workers at the median price of 35 to 40 r/kg. In the opinion of the director of this plant there was little point in studying new techniques to make the meat of a higher quality (and a higher price) rivalling imported sausage since then they would no longer be able to barter the same volume of a 'cheap' but useful commodity to their unpaid workers. The other main processor in Tura, the Tura Aviation Enterprise, had a similar philosophy. Through their special access to parts of the District the Enterprise has established an enviable fresh supply of meat, which in the conditions created by the new cashless economy has led them to invest in their own smoking and vacuum-packing technology. This small plant is used primarily to feed 'their own' workers (approximately five hundred individuals) with processed meat 'sold' against back-pay. The young director of this enterprise was practical in suggesting that if they had 'extra' they might sell it to outsiders, but at the present time they would rather 'feed their own first'.

The local market structure of Tura, and perhaps many other places in Siberia, is a set of nested paradoxes much like in the example of consumption in the name of science which David Ellis (this collection) observed in Papua New Guinea. Overall, there is both a great social need and a high demand for fresh, high-quality, locally produced reindeer and caribou products. However, given that no one can be paid a cash wage for bringing production to market, there is a strong social pressure to keep production confined to tightly defined social networks and to limit the price 'charged' for these bartered products as a favour to those within those networks. The combined set of these networks practically 'saturates' the regional demand, with the significant exception of families who are pensioned or otherwise 'unemployed' ('unattached') to one of the major employers. Thus one can argue that a viable 'market' exists only in a very abstract, theoretical sense in the region and that it is practically untapped in the major cities of Noril'sk, Igarka and Krasnoiarsk. The reality of the trade in caribou and reindeer meat is that it is embedded within a set of competing social networks, no one of which will be permitted by the others to develop a larger volume and hence a monopoly on meeting people's needs. The only way to build a higher volume would be to find a source of credit and a form of organisation to immediately link dispersed local hunters to urban Russian markets thousands of kilometres away. Until these conditions are created, it is far preferable, or at least less costly in terms of energy and resources, to import Brazilian chicken and New Zealand lamb, despite the loss of income to what an outsider would abstractly imagine as a coherent local market.

Can markets ever be situated?

For anthropology, these examples of competing interests within a complex network of nested barter markets allow us to question the topographical metaphor of 'markets' in general. The market *relationship* is very poorly served by the image of a market as if it were a circumscribed and established space where one could instantly conduct a reasoned negotiation of exchange (Hann 1998). On first principles, there is a lack of consensus about which social agents have the moral right to sell food – even when the threat of malnutrition lurks in the background. In the Soviet period, basic foodstuffs like caribou meat were supplied at subsidised prices which were merely symbolic. Furthermore, in this example, perhaps more strongly than in examples from other places, there is a remarkable plural aspect to the idea of a 'public' interest. Members of the public can be divided ethnically, in terms of federal divisions, and in terms of professional loyalty. Often it seems that the hold of various professional associations ('collectives') is much stronger than any national or regional interest. From a 'Western' point of view this seems perplexing. Obviously there already exists a demand, a skill base, and the technology to develop a vibrant regional economy independent of central state legislation. However, the fact is that the main entrance to the topologically situated 'market' is jealously guarded by various social groups. Thus, if it were possible to identify a place where food could be sold, it is not clear if it should be sold at all, and if it were, it is not clear who should be doing the selling. For development agencies, for whom this research was commissioned, the clever trick becomes designing training programmes and technological exchanges such that no matter which group gains control of the entrance to the market it can nevertheless service a wide base of the population almost in spite of its own interests.

For anthropology, however, our task seems to be to discover a way to develop a language which first replaces the topologically tidy metaphor of the 'market' and then evocatively captures the range of social and political interests which condition the economic. One way forward, as my co-author has suggested, is to experiment with new ways of combining ethnic and ecological metaphors with an analysis of economy. This particular project suggests that metaphors related to domestication ('wild', 'feral', 'domestic') and kinship ('our own' versus 'other') offer promising alternatives in this direction.

Setting aside for the moment the literary devices used in this chapter to make Siberian markets seem paradoxical to Euro-American readers, it should be mentioned that this political ecological approach to describing exchange relationships is used by local producers themselves when they describe ways of improving their way of life. Indeed, the way forward, suggested by hunters themselves is in the formation of new, collectivist institutions which use kinship as an idiom in which to build trust and to control distribution.

The majority of the 'producers' who spend long nights out on the land harvesting migratory caribou or herding reindeer currently organise themselves into the so-called clan communes (*rodovye obshchiny*) which are the focus of so much administrative effort and ethnosociological research (Popkov 1994, Sirina 1999, Fondahl 2000, Gray 2000). The clan communes are groups of Evenki and Yakut people, some of whom might be related, who have taken out long-term leases on specific parcels of land in order to concentrate upon fur-trapping and caribou harvesting without the social guarantees or baroque division of labour of state farms. In the early 1990s the formation of these ethnically and kinship-stratified organisations was seen as the cutting edge in reform policy aimed to 'give the taiga back to the Evenkis'. It was a pioneering experiment in autochthonous development wherein primary producers were gallantly evicted from the state-socialist-planned economy reducing, as it were, the number of competitors for those who knew how to monopolise exchange networks.

In 1999 in the Evenki district there were still over thirty registered clan communities, the majority in various shades of insolvency, who were nevertheless able to make deliveries of fur, fish and meat. However, they were generally forced to accept whatever price that helicopter pilots, oil-workers and store-owners offered them for their wares. In the views of some of these producers, the best way for them to achieve better access to hungry consumers would be for them to vertically integrate their ethnically distinct clan communities into larger organisations. The precise forms of these proposed enterprises are quite varied. They range from alliances with ethnically distinct trading posts (*faktoriia*) which would serve as a form of consumer cooperative able to bargain on a greater scale with outside institutions, to a type of 'community development corporation' (*posel'kovie unitarnoe predpri-*

iatie) which would employ its state-derived tax-free status to coordinate the optimal distribution of goods. These formations differ from those of the recently jettisoned collective and state farms in one important aspect – the right to participate would be restricted to members of 'sparse aboriginal peoples of the north'.

That the aboriginal dimension to these socio-economic organisations has become an issue is in my view one of the most significant developments in the history of this region. There is a strong argument to be made that social relationships in Soviet Siberia, unlike in other parts of the world, were negotiated without much recourse to the idiom of aboriginality (Anderson, in press). Partly under influence from indigenous non-governmental organisations like the Inuit Circumpolar Conference, and partly due to the negative reaction to the fact that the newly deregulated exchange mechanisms came to be so completely monopolised by people perceived to be outsiders, the majority of rural Siberian hunters and herders today feel that some sort of exclusive corporation is needed 'just for Evenkis' or 'just for Yakuts'. Although entirely understandable, this development is nonetheless tragic if one reflects on the nuanced way that kinship and profession were applied in the Soviet period to negotiate more complex (and much more beneficial) relationships with settlers (Anderson 2000a).

In order for the kinship-stratified idiom of the present clan communities to be consolidated into an ethnically stratified corporation, most actors are waiting for some definitive action from the Russian state. Although overtures in this direction have been made with the passage of a federal law in 2000 defining the rights of sparse native peoples of Siberia, a set of draft laws on reindeer herding, and a colourful collection of regional statutes, the most concrete action seems to be being taken in a sphere controlled by agents even more distant than the settlers who guard the Siberian markets. Environmental organisations such as the World Wildlife Fund for Nature and the Global Ecological Facility, as well as various national development agencies, have been vocal in proposing that large portions of taiga and tundra be set aside for the 'environmentally' inspired ends of preserving migratory bird and caribou populations and ensuring the 'sustainable' development of indigenous peoples (Koester 2000). The motives of these agencies are varied, and cannot be explored in any detail here. Suffice it to say that various projects tabled or in progress for central Siberia in Taimyr, Evenkiia and Turukhansk county involve altering the juridical status of large swaths of territory on the basis of only cursory consultation with local hunters and herders. Given the very complex way that Euro-American ideas of discrete markets have been applied to this vociferous social setting, it is a somewhat alarming thought to consider how the less abstractly defined idea of a 'protected territory' might be operationalised in a region where excluding access to space and to resources and services has been a principle of social power for at least a century.

Returning to our analysis of the ecology of economic underprivilege, it should not be at all surprising that the ideas of inscribed markets have been taken in a direction that is different from our preconceptions of markets as places where anonymous strangers meet. Instead we can clearly see a set of ideas featuring domestication, kinship and ethnicity wherein markets are not spatially circumscribed. Rather, human relationships are ordered or adjusted to work within limits which are not purely territorial in nature. The value of ethnography comes in challenging development economists to imagine a different way of describing market relationships than that suggested by a definition of a market as a discrete place where capital and skills come together to happily produce goods for consumers. Instead, there should be a way of describing an ecology of relationships that protects and nurtures local interests rather than throwing them open to opportunistic predation.

Notes

1. In this chapter I use the Canadian word 'caribou' to refer to migratory populations of wild *Rangifer tarandus* more commonly known in the literature by the folk name 'wild reindeer'. I reserve the term reindeer for domesticated *Rangifer* only. I have changed these terms to avoid the repeated clumsy use of 'wild' and 'domestic' since one major theme is the transition of one type of *Rangifer* to another (and back again).
2. The price of fuel tripled in the autumn of 1999.

10

CONTRASTING LANDSCAPES, CONFLICTING ONTOLOGIES
Assessing environmental conservation on Palawan Island
(The Philippines)

Dario Novellino

As a result of a flawed appeal to universal human values by conservationists, debate on environmental protection is trapped in Western categories with detrimental implications for the lives and well-being of indigenous communities. Recently the Philippine government has embraced a new discourse of caring environmentalism, much influenced by Western/Northern conceits, where these problematic categories are only strengthened and even given legal standing. Cosmologies like that of the Batak of Palawan Island can be profitably examined to help us escape these impasses of eco-discourse (see Figure 5). Specifically, they shed light on the false dichotomy between technocentric and ecocentric environmentalism, where the technocentric variant is understood as optimistic but arrogant towards nature, and the ecocentric is generally characterised as pessimistic and romantic (O'Riordan 1981). Batak cosmology does not fall into this trap.

One does not need to look far to understand some of the principles underlying the basic tenets of the conservationists' 'faith'. For instance, the well-known slogan 'think globally and act locally' appears to be based on the assumption that the whole of humanity strives for a common goal, that is the protection of the world's natural system. Of course, I am not disagreeing with the noble objective of saving the Earth. Here, I am questioning the epistemological status of notions such as 'respect' and 'responsibility for' the environment regarded by conservationists as universal, and thus applicable cross-culturally.

The ethnography I present here suggests the Batak[1] have always been thinking 'globally', in the sense that they normally draw a causal connection between certain actions (over-hunting, over-harvesting, improper sexual behaviour, etc.) and the impact that these may have on the world (see Cosgrove 1993). Prima facie, attributing responsibility for the fate of the world to humans seems to be endorsed by both conservationists and Batak. However, what differs is the perception of 'globality', how the cosmos is imagined and, indeed, the way in which the relation between causes (human actions) and effects (ecological consequences) is apprehended vis à vis the role of non-human subjectivities like animals, plants, other entities. Unlike conservationists, Batak do not project into the world a generalised fear that humanity's destructive power can bring life, as we know it, to an end, and that 'culture' can take over 'nature'. Rather, they are more concerned with how socialisation between human and non-human agents come into being, and about the repercussions (both positive and negative) that the latter may have on the world.

This chapter shows that Batak cosmology presupposes that humans, animals and other non-human agents possess an 'intangible essence' (*kiaruwa'*), the source of thought and action. This introduces a distinctive perspective of the relation between Self and World, one where the common denominator of all beings is not nature (physical substance) (cf. Descola 1986, Viveiros de Castro 1998)[2] but rather, this 'intangible essence', because of which living entities are, and behave, as autonomous subjects. This is a reversal of Western cosmology where nature (substance) enjoys a status of ontological supremacy and universality. For Batak, to be and act as subject is not the sole prerogative of human beings, but is rather shared by non-human agents holding distinctive points of view about the world, what I call co-apprehension. Batak too, like conservationists, have a 'global' perception, but one where social relations, not nature or biodiversity are its fundamental constitutive elements.

Furthermore, this paper argues that conservationists' tendency to define local environmental problems as global problems, and to propose internationally concerted managerial solutions to 'regional' ecological issues (e.g., zoning of protected areas on Palawan Island),[3] has the effect of ideologically disempowering indigenous communities, while jeopardising their livelihood patterns. To support my argument, I contrast Batak cosmological principles and perceptions of landscape with the presuppositions underlying the classification and implementation of protected areas systems. I first review Batak ways of relating to non-human agents, giving particular attention to the role of the shaman in the 'healing of the world' and in the re-distribution of resources. I then discuss the way in which environmental laws in Palawan rationalise the landscape. Finally, I elaborate on the position that in Batak ideology, *kiaruwa'* – the source of will, self-assertion and desire – is equally possessed by humans, animals and other non-human agents. Again, this contrasts with Western cosmologies which 'postulate a physical continuity and a metaphysical discontinuity between humans and animals' (Viveiros De Castro 1998: 479), where humans are considered unique in possessing an intangible essence (the soul), and thus superior to other beings. Batak regard *kiaruwa'* as the source of intentionality, power of agency and a point of view. It follows that it is at the heart of human-non-human interactions.

This observation introduces an ontological argument which takes the form of asserting that there is no universal human ethic about the use and management of the environment. Here, the key point is that metaphysical presuppositions underlying Batak understanding of ecological imbalance, species decline, and the role of humans in saving the planet are based on culturally specific notions of interaction between humans and non-humans, whose incorporation in global environmental discourse would be untenable or at least problematic. Batak do not regard environmental degradation as an object of managerial solutions *per se*. Emphasis is placed, instead, on the social dynamics which have lead to environmental damage. This is blamed not so much on peoples' inadequate technologies or destructive subsistence practices, but rather on their incapacity to maintain appropriate relationships with non-human agents and among themselves.

Batak cosmology and perception of landscape

Conscious agents, human and non-human

The Batak live in a tropical forest environment. The forest encompasses the village, with its plants, animals, rocks, rivers, patches of fields under fallow, but it is also inhabited by *panya'en*, entities and super-human beings which may dwell in trees, rocks, water pools, etc. Certain trees are embedded with social relations through genealogies and settlements' histories. The Batak's close interaction with a tree-covered landscape 'imposes a reorganisation of sensibility' (cf. Gell 1996) and emphasises the fact that experience is not only visual, but the environment is also apprehended through olfactory, auditory and other senses. For instance, among the Batak it is not the visual appearance of a stranger that can make a child sick, but rather his smell (as of perspiration) to which the child is not accustomed. In a similar way, malevolent non-human agents can be attracted by the smell of newcomers and react accordingly.

The forest environment not only affords seeing and smelling, but also listening. Monkeys and birds are generally heard before being spotted and chased. Understanding and interpreting the sound of the forest is perceived by the people as a prerequisite for survival. According to my Batak informants, the whistle of the *sagguai* bird, or the sound of the gecko, are the most common warning signs informing the people about the presence of a *panya'en*. Ignoring such signs can lead to illness or even death. Overall, while walking in the forest, a Batak is not only aware of his own perception of the world, but also of how his locomotion, smell and the sound he makes may be perceived by other living entities sharing that same landscape.

For the Batak, procuring certain resources not only depends on their physical availability, but also on the way they provide affordances to humans. Specifically, affordances or capacities for (see Gibson 1979) are strictly linked with the quality of the relationships that people are able to establish with their environment, with the masters of animals, plants and other non-human entities. To cite an example, during collective fish-stunning with a poisonous vine, a specific elder is in charge of coordinating all members of the fishing group (Novellino 1997, 1999a). Then, super-human beings and various entities are invited to take part in the fishing, as well as the afore-mentioned *panya'en* who dwell in trees, rocks etc. The elder invites all men to participate in a group discussion. That same night, the shaman, accompanied by his assistants, will spend several hours on the river shores, chanting and falling into trance. During his 'out-of-body' experience, the shaman claims to be able to see the quantity and the type of fish in the river. The shamans' landscape coincides with, but also transcends the tangible landscape. The *kiaruwa'* of the shaman can locate precious resources that the naked eye cannot detect. It flies and sees the world from above, or dives into rivers and sees the world from below. He may share a vantage point with birds and fish, and thus perceive the landscape from different perspectives. He shares these visions with the participants and admonishes them not to break certain prohibitions.

This introduces a concept of alterity where the social attributes of many living beings are those of *kiaruwa'* as this interacts with other conscious agents (cf. Lima 1999). Thus human experience also is characterised by complex interactions between different 'subjectivities', which apprehend the common environment in different ways through their distinctive physical bodies and *kiaruwa'*.

Before describing shamans' practices for re-establishing the social/cosmic balance, I will provide a brief outline of Batak views of the cosmos. The Batak envision a cosmos of seven layers consisting of a central tier, inhabited by humans, animals, plants, super-human beings and aggressive entities, layered between an upper world and a lower world. There is a cosmic axis crossing the seven layers, around which the world rotates. The sun shines through the different layers, and when it lights up in the lowest level, the rest of the world is in the dark. The lower world is 'the land of the dead', a sort of purgatory, where 'dead people' lead lives resembling those of the living. After an unspecified period, their *kiaruwa'* will ascend to the seventh upper level, where the creator God dwells. Stars and the moon are imagined to be steady in the sky. Often, stars are said to be the root tips of a shining tree growing in the seventh upper level. The central layer is believed to be shaped as a large disk, surrounded by sea and sky. *Puyus*, the highest mountain in 'Batak land', is regarded as the original place of all malevolent *panya'en*. The upper layer of the universe is the domain of other immortal super-human beings. The *gunay gunay*, at the edge of the universe, is the place of origin of the master of rice, the female deity *Bay Bay*, and that of the master of bees, *Ungaw*, who are believed to be husband and wife.

'Correct' use of the common environment

Now let us return to the significance of shamanic practices. Not only fish-stunning, but also other activities such as hunting, the clearing of new fields, and honey gathering are accompanied or preceded by divination. Like all animals, bees are imagined to have their own master, *Ungaw*, a 'mystical' beekeeper in charge of their welfare. Batak envisage a kind of cyclical system in which the seasonal production of honey and rice depends upon the flow of bees and of the *kiaruwa'* of rice from the *gunay gunay*. Bees need to be explicitly requested by the people during a ceremony, and they must be well-received.

The continuity of cooperative behaviour and reciprocal exchange, the maintenance of good relationships with the other entities, are all associated by the people to what is perceived as a 'correct' use of the common environment shared with animals, plants and non-human agents. In contrast, over-hunting, over-harvesting of plants and animals, and wasting the meat of killed game are said to offend the keepers of animals and certain *panya'en*. In this respect, the role of the shaman (*babalian*) is of vital importance. He helps to mediate between the tangible and intangible landscapes, he works to restore relationships between humans and other entities and to maintain the social/cosmic order, hence ensuring access to natural resources (Novellino 1997, Novellino 1999a).

Particularly relevant to shamanic practices of 'healing the world' is the *burungan*. It is believed to be located in the central layer of the universe, and it is also defined as navel of the sea or as root or base of the sea, from which water is expelled into the world. When storms and floods ravage the world, this is because the *burungan* is 'open' and thus water flows out freely and abundantly. Two enormous stones are believed to regulate the flow of water at the *burungan* entrance. The more such stones are distant from each other, the more intense the flow of water will be. When the stones touch each other, the *burungan* is said to be 'closed', and the world is blessed with sunshine and good weather. According to my shaman informants, a chicken 'big like a house' rests on a huge metal bar on top of the *burungan* opening, and an enormous dragon, known as *tandayag*, dwells in the deep sea. When incest and other improper sexual behaviours are committed, the *tandayag* dragon will wake up, beginning to shake; then, if no proper rituals are performed, it will initiate its search for the incestuous couple. The shaking of the *tandayag* will cause the water to overflow from the *burungan* and to reach the terrestrial portion of the world in the form of flood. The outflow of the water will make the metal bar wet and, as a result, the rooster will slip off the bar, remaining attached by one claw only. Losing its balance, the animal will begin to flap its wings. The shamans claim that should the rooster slip off the bar completely, the entire universe will collapse and disappear. The wind produced by the flapping of wings is said to be the source of storms damaging crops and village huts and causing dead branches to fall on the ground.

This particular condition of the Earth requires shamanic intervention, and a specific ritual is performed in order to heal and 'to renew' the world so that the sun will shine again. Because ecological imbalances are also attributed to the infringement of customary norms, phenomena such as storms and floods may be perceived as the outcomes of social problems, whose direct repercussions are observable in the environment. A practical example will illustrate my point.

In August 1999, while I was on a visit to the Batak of Tanabag, the inclement weather was attributed by the village shaman to human misbehaviour in general, and particularly to a severe case of incest which had occurred in the community. A shaman had eloped with the wife of his son, who had died of tuberculosis the year before. Their union was considered illicit because their case had not been evaluated by the 'council of elders', neither had they paid the appropriate tribute or offering to the *tandayag* dragon. According to Batak, in case of incest the tribute should consist of a large offering of plates, money, chickens, and the blood of the incestuous couple.

To then 'renew the world', the most important phase of the ritual is the trance performed by the shaman. The shaman holds coconut oil in one hand, while dancing. During the trance the *kiaruwa'* of the shaman will be approached by selected spirit guides, and will require their help to reach the *burungan*. These spirit guides are associated with animal species, and they represent their 'spiritualised' version. They are the *kiaruwa'* of animals 'of the water' and 'of the higher up', and include the swallow, the otter, the monitor lizard and the river turtle. The *kiaruwa'* of the river turtle is considered the most enduring and capable of confronting the fury of the water at the *burungan*. It will also play a shamanic role by dancing the same dance as the shaman, thus fostering the closing of the two boulders over the *burungan*. While the turtle's *kiaruwa'* dances, the shaman smears the two boulders with coconut oil, facilitating the coming together of the two stones above the *burungan* opening, thus stopping the water outflow. According to the informants, a particular malevolent *panya'en* appears at the *burungan* site in the form of an attractive woman, and she will try to call on her the shaman's attention. It is said that if the shaman looks back at the malevolent *panya'en*, he may be hit by the water outflow, and fall inside the *burungan* hole. Finally, with the assistance of the spirit guides, the shaman will try to place the rooster claw back on the metal bar. In so doing, the rooster will stop flapping its wings and the storm will end.

Kiaruwa' and relationships between human and non-human agents

Now, let us explore the Batak notion of *kiaruwa'*, and some crucial aspects of the perceived relationship between human and non-human agents. Elsewhere (Novellino 1999e) I have defined *kiaruwa'* as 'life force', the vital principle associated with breath. However, I find this definition too narrow since it fails to convey the complex meanings that Batak attribute to this notion. *Kiaruwa'* is, in fact, not only the vital principle but the source of individual will and self-assertion. When it is separated from the body, the kiaruwa' of a human retains the 'character' and 'quality' of the person. According to the Batak, the *kiaruwa'* is believed to enter and fill the body through the whorl of the hair (the region of the fontanelle). It is compared to a knife entering its sheath, so that a *kiaruwa'* of a certain species can only be contained within the body of that species (i.e., that of a human cannot enter an animal body and vice versa). Also, animals as well as rice (because of its human origin) are said to possess *kiaruwa'* though the qualities it confers vary qualitatively between humans and animals.[4] In relation to humans, *kiaruwa'* is also the focal point of illness aetiology and curative treatments. Temporary departures of the *kiaruwa'* can cause lassitude or sickness. Death coincides with a longer separation of the *kiaruwa'* from the physical body. The separation of the *kiaruwa'* is a threat to the integrity of the 'self' (cf. Roseman 1991); on the other hand, it allows humans to establish contacts with the intangible world.

In Batak society, animals and also *panya'en* entities are mainly classified by the habitats they share – land, fresh water, sea, up high, people's houses, etc. Communication between humans and benevolent *panya'en* (e.g., the master of bees, the master of rice) is conceived as possible by virtue of the fact that they are all regarded as 'persons' (*taw*), and their *kiaruwa'* is the source of a 'human way of thinking'.

Out of the trance and dream experiences, humans will attempt to affect animal behaviour through the use of specific techno-symbolic devices. A large number of such devices are employed, for example, in connection with honey-gathering where the plant *eya' eya'* (*Mimosa pudica* Linné) is employed to reduce bees' aggressiveness. The informants explain that bees will become weak like the *eya' eya'* leaves kept by the gatherer. *Mimosa pudica* is a sensitive plant; when touched, the leaflets immediately fold together upward and the main stalk folds down, giving the impression that it is losing strength. Batak also claim to be able to direct swarms of bees from one place to another. To this purpose, they use *suway suway*, which are wooden sticks to which a fibre is tied to form a semicircle, resembling the shape of the honeycomb. According to informants, the way in which *suway suway* are positioned and inserted into the ground will provide indications to the master of bees of where he should disperse his children (the bees). The utilisation of *eya' eya'*, *suway suway*, and similar tool-signs is also based on the principle that certain actions, or the utilisation of natural or human-made objects may produce desired outcomes, due to their analogy, affinity or opposition to something else. These tool-signs may be perceived by Batak as forms of communication between humans and super-human beings, for instance, the master of bees.

The human ability to communicate with animals and other entities is fully expressed when a shaman undergoes an 'out-of-body' experience. Behind their physical appearance (the body) and, at the level of the *kiaruwa*', both human and certain non-human agents can simultaneously apprehend different realms, and thus share similar points of view on what each realm affords. When the *kiaruwa*' is separated from the body, possibilities for co-apprehension increase, and humans can move and enter dialogue with the Masters of Animals and other benevolent *panya'en* across different ontological realms. It is important to point out that co-apprehension, or mutual intelligibility between different species and entities, is also perceived by Batak as possible by virtue of their common 'human origin'. Batak have several myths narrating how certain animals (e.g., a species of black snake) and plants (rice and other crops) have experienced the condition of being a human, or better, are 'ex-humans' (Viveiros de Castro 1998).

Having drawn out these points, I should make clear that Batak cosmological principles and their idealised relationship with non-human beings is not necessarily reflected in peoples' daily behaviour towards animals and plants. In other words, cosmologies and myths are not the sole framework for imbuing experience with meaning, and determining the way in which all actions are carried out (cf. Vayda 1996). Batak, in fact, do not seem to place any particular emphasis on moral obligations towards other species. Only the killing of pets (dogs and cats, to the exclusion of chickens) is regarded by the people as a bad custom. Remarkably, it seems that Batak do not show restraint from hunting certain species, such as the hornbill whose numbers are visibly diminishing, and which is thus particularly vulnerable to local extinction. Flying squirrels (see Figure 8) are killed by smoking their holes in tree trunks, causing the death of both adults and very young specimens. *Anibung* palms (*Oncosperma* sp), ten to fifteen years old, are felled to extract a couple of kilograms of edible bud heart, just enough for one meal (Novellino 1999d). And river turtles, in spite of their position in myths and cosmology, are subjected to the most horrifying death. According to the Batak, the most common way of killing the turtle is to drop it alive into the fire (Novellino 1999c).

Undoubtedly, from the standpoint of conservationist ideology, Batak practices do not reflect the romanticised idea of the noble savage living in harmony with nature (Novellino 1998). Batak do not practice conservation for its own sake, neither are they interested in the protection of single species. Moreover, the environment is not imagined by them as something which is sustained through the ethically sound behaviour of a superior species, but rather through the maintenance of social relationships between 'species-bound identities' (Howell 1996).

Let me provide a practical example to further strengthen my argument. In 1989 a group of indigenous people from Pinatubo area (Zambales, Northern Philippines) was relocated through the assistance of a German citizen to the Batak territory. For the first time Batak were confronted with a very unusual situation: another group of indigenous people was blatantly violating customary food taboos in their own territory. In fact, these people were capturing snakes and frogs, consuming them as food. It is interesting to note that the fact that they did not die from eating such food (which the Batak regard as poisonous) did not modify people's perceptions about these animals. What is relevant here is that Batak were neither interested nor curious to witness the possible effects of frog and snake meat on human health. On the contrary, they were concerned that the breach of food taboos by the newcomers would have angered a number of *panya'en*, with possible and dangerous repercussions on Batak, and their environment as a whole. This is to say that eating customarily inappropriate food, over-hunting and over-harvesting are perceived by Batak more as asocial behaviour than as ecologically unsound activities.

Environmental laws: The landscape of non-Batak others

I now wish to turn attention to another 'cosmology', that of the Filipino environmental laws inspired by the World Conservation Union (IUCN). The analysis of such laws indicates that Western projects to conserve natural habitats and indigenous perceptions of the environment do not stand in some taken-for-granted relation one to another. On the other hand, government options for environmental conservation such as the ban on shifting cultivation and the establishment of national parks are based on a poor understanding of indigenous practices and worldviews. As a result, such options can produce grotesque and rather dangerous outcomes for the survival of local communities.

I need to point out that, in Palawan, conservation is influenced by both ecocentric and technocentric approaches to environmentalism, to the extent that it is often impossible to make a neat distinction between the two. Supposedly ecocentric and romantic ideas, for instance seeing the Batak as closer to nature and thus perhaps less fully human, easily coexist with more technocentric projects such as zoning the landscape.

The National Integrated Protected Areas System (NIPAS) and the Republic Act 7611, also known as the Strategic Environmental Plan (SEP) are relatively new laws, enacted in February and June 1992 respectively. Such laws establish the legal basis for the protection and management of

Figure 8 A Batak with his
catch: a flying squirrel. *Photo:*
Dario Novellino.

the environment in Palawan (SEP), and nationwide (NIPAS). Protective
measures proposed by the laws include the demarcation of areas either as
off-limits to the human population, or reserved for local 'indigenous cultural communities' (ICC),[5] or both (Novellino 1998, 2000a, 2000b). As
illustrated for instance by Anderson and Nygren (this volume), such categories of protected areas have also been implemented elsewhere. Local
communities are expected to limit or refrain from certain subsistence activities when their territory becomes divided into management zones with
different levels of protection (from strictly non-touchable to controlled
use).

As established by the NIPAS (Republic of the Philippines 1992a), total
protection is likely to be enforced in areas defined with the following categories:

- Strict Nature Reserve: 'possessing some outstanding ecosystem, features
and/or species of flora and fauna of national importance';

- Natural Park: 'relatively large areas not materially altered by human activity
where extractive resource uses are not allowed';

- National Park: 'forest reservation essentially of natural wilderness character which has been withdrawn from settlement, occupancy or any form of exploitation';

- Wildlife Sanctuary: 'areas which assure the natural conditions to protect nationally significant species, groups of species, biotic communities'.

On the other hand, human occupancy and resource utilisation are contemplated in categories such as:

- Protected Landscape/Seascape: 'areas of national significance, which are characterised by the harmonious interaction of man and land';

- Natural Biotic Area: 'an area set aside to allow the way of life of societies living in harmony with the environment to adopt the modern technology at their pace'.

As can be easily anticipated, conservation measures based on land zoning, such as those provided by the NIPAS, disintegrate the unity of the indigenous country. Land categories, by being cut off from the whole, become meaningless to the Batak. People do not perceive their country as an enclosed and atemporal island, but rather as a continuum of indivisible features (Ingold 1986) which are the repository of previous experiences, past events and social relationships (Rosaldo 1986). In NIPAS law, indigenous communities represent one of the patchwork pieces to be placed in the most appropriate spot (a Natural Biotic Area), provided they continue to live 'in harmony with the environment' (Republic of the Philippines 1992a:3). In other words 'the harmonious interaction of man and land' (Republic of the Philippines 1992a: 3) is almost viewed as a precondition for residing in 'areas of national significance', such as Protected Landscape/Seascapes (Republic of the Philippines 1992a: 3). Undoubtedly, a policy such as NIPAS, treating both indigenous communities and wildlife as species which need protection, is far from being innocent. It is rather, a political act to 'ontologise' cultures, to assign a different existence to indigenous populations. This has the effect of removing the people from the space they occupy (Fabian 1983), thus depriving them of agency and history. In a similar vein, in this volume, Ellis has neatly demonstrated how conservation biologists negate both local history and subsistence needs (see also Adams and Chatty). Following a parallel line of argument Halder (this volume) has placed additional emphasis on the incommensurability between Lakota Sioux and official perceptions of sacred places.

The same criticism of NIPAS applies as well to the Strategic Environmental Plan (SEP) for Palawan, and to projects pursuing similar objectives (e.g., the European Union-financed Palawan Tropical Forestry Protection Programme). The SEP law, also known as the Republic Act (R.A.) No. 7611, provides a comprehensive framework for sustainable development and contains a package of strategies on how to prevent further environmental degradation. The centrepiece of the strategy is the establishment of the Ecologically Critical Areas Network (ECAN), which places most of the province under controlled development. The areas covered by ECAN include three major components: terrestrial, coastal/marine and tribal ancestral lands. Core zones are defined as areas of maximum protection and consist basically of steep slopes, first growth forests, areas above one thousand metres elevation, mountain peaks, and habitats of endemic and rare species. The law establishes that core zones 'shall be fully and strictly protected and maintained free of human disruption ... exceptions, however, may be granted to traditional uses of tribal communities of these areas, for minimal and soft impact gathering of forest species for ceremonial and medicinal purposes' (Republic of the Philippines 1992b: 101). Interestingly enough, the ECAN core zone coincides with significant portions of the Batak hunting and gathering ground. For instance, the resin of *Agathis* trees is usually extracted in commercial quantities around one thousand metres above sea level. In addition, several game animals, especially flying squirrels, are trapped for food around these altitudes. A similar situation was identi- fied by Ellis (this volume), who claims that most of the environments defined by biologists as untouched by humans are in fact anthropogenic landscape *par excellence*.

Buffer zones represent the most elaborate component of the ECAN and are designed to serve a multiplicity of purposes. According to R.A. 7611, they fall into three categories known as restricted use areas, controlled use areas, and traditional use areas where 'management and control shall be carried out with the other supporting programs of the SEP' (Republic of the Philippines 1992b: 101). The only area within the so-called terrestrial component which mentions agricultural practices is the multiple/manipu- lative use zone: 'areas where the landscape has been modified for different forms of land use such as extensive timber extraction, grazing and pastures, agriculture and infrastructure development' (Republic of the Philippines 1992b: 101). It is crucial to point out that a large number of indigenous communities are occupying marginal upland areas, which fall under the wider definition of buffer zones. Furthermore, the law never mentions indigenous swidden framing practices, hence we may easily come to the conclusion that only imported methods such as terracing and hillside farm- ing will be allowed in multiple/manipulative zones. There is no specific indication of where such zones are located, but it is legitimate to anticipate that these areas are occupied by a vast majority of migrants, rather than by 'traditional' indigenous communities.

It is important to point out that Section 11 of R.A. 7611 includes tribal ancestral lands among its categories. The law specifies that 'these areas, tra- ditionally occupied by cultural minorities, comprise both land and sea areas identified in consultation with tribal communities concerned and the appropriate agencies of government' (Republic of the Philippines 1992b:

100). It is frustrating to learn that even tribal ancestral lands 'shall be treated in the same graded system of control and prohibitions except for stronger emphasis on cultural consideration' (Republic of the Philippines 1992b:100). At the same time, we are assured that 'the SEP ... shall define a special kind of zoning to fulfil the material and cultural needs of the tribes, using consultative processes and cultural mapping of the ancestral lands' (Republic of The Philippines 1992b: 100). It is clear that SEP, with a high degree of naïvety, proposes the protection of indigenous culture on the one hand, and the implementation of Western zoning criteria in tribal lands on the other. So far, the law and its promoters have been unable to provide a convincing argument of how this can be achieved.

As of now, in Palawan, the zoning of protected areas is still in the process of being implemented and finalised. Presently, government teams composed by foresters, Filipino scientists, and local guides are surveying the area, carrying out biodiversity assessments, measuring elevations, demarcating boundaries, marking trees and boulders. However, no evidence is yet available on what impact the actual implementation of ECAN (Ecologically Critical Area Network) will have on residents of the region. However, one can almost predict tragedy. This is suggested by an examination of the outcomes of a local government ban on swidden cultivation, and the way in which local Batak and Tagbanuwa communities have been excluded from the management of the St. Paul Subterranean National Park. In the first case, the ban has irredeemably altered the whole indigenous agricultural system. Secondly, it has also affected the genetic diversity of cultivated plants and local crop varieties are becoming rare and even extinct. Ultimately, the prohibition is placing an insupportable burden on the surrounding forest. In fact, to compensate for the loss of agricultural products, Batak have been forced to over exploit their own resources (e.g., the resin of *Agathis* trees, rattan, and honey). As far as concerning the 'St Paul Park', the hands-off style of strict protection has curtailed the customary food-seeking practices of the affected Batak and Tagbanuwa communities. (Similar degrees of environmental injustice seem to be taking place in the biological reserve of Indio-Maíz and in the Lakota Black Hills, as Nygren and Halder describe in this volume).

In 1991, one Batak named Paya Paya, and two Tagbanuwa were arrested and detained for allegedly having burned primary forest inside park boundaries. Later investigations showed that the area utilised for agriculture by the three individuals was located outside the boundaries of the park, and consisted of secondary and scrub vegetation.

Batak often talk amongst themselves about government ordinances and laws imposing limitations on the use and access to forest resources. I recorded one of such spontaneous discussions in August 1999, in the settlement of Kalakuasan. Catalino, a Batak in his thirties stated, 'If you come close to the government, the government says "You must learn this ... you must learn that … so that you can live well". The more you come close to the government to get rid of your poverty, the more the government squeezes you, making you poorer and poorer'. Padaw, the local shaman remarked: 'You are right! They squeeze us by prohibiting everything. The Government claims that it wants to help us, in reality they even forbid our people from planting rice … They ask us to sign agreements to define what has always been ours, from the time of our ancestors!' As external pressures increase, Batak are left with no other alternative but to come to terms with a number of new and increasingly complex issues such as environmental laws, whose rationale is not immediately comprehensible to them.

Evaluating Batak and conservationist 'cosmologies'

At this point in my argument, the reader may ironically be tempted to compare Batak view of a layered universe to the way that environmental discourse makes layers on landscape. However, from another perspective, I contend that Batak cosmology is similar neither in structure nor content to the conservation logic of zoning. Firstly, Batak do not perceive landscape and the 'cosmos' as a *tabula rasa* which can be inscribed, measured, and experienced accordingly. As I have previously suggested, Batak perceive landscape (the *burungan*, the universe with its multiple layers, etc.) as it simultaneously occurs to them and to the other living beings with whom humans share overlapping ontological domains. Thus, it is 'the point of view of the others' which also serves to define one's own position in the world. Conversely, zoning creates an ego-centred and perspectival landscape of views and vistas, where it is only the expert's perspective from which seeing occurs (Bender 1993). Secondly, in environmentalist discourse, the zoning system, through its categories of protected areas, tends to introduce an anonymised and ahistorical image of the ideal relationship between humanity and nature. Here, I want to insist that Batak landscapes and cosmological layers are perceived by the people as having histories, but 'nature', epitomised in protected areas, does not.

Another crucial argument that I have advanced in this article is that in Batak understandings of the interaction between human and non-human, it is not the 'intangible essence' (*kiaruwa*'), but rather the body which represents the 'great differentiator' (Viveiros de Castro 1998). What then does Batak spiritual undifferentiation between humans and certain non-humans tell us about the ontological supremacy and universality of nature, envisaged in much of contemporary *conservationist* discourse? To begin with, the Batak worldview challenges the naturalist perspective, typical of Western cosmologies, which supposes that 'the nature/society interface is natural … [and that] social relations … can only exist internal to human society' (Viveiros De Castro 1998: 473). If in conservationist discourse human societies are modelled after an idealised notion of 'nature', in Batak cosmology the environment is modelled after an idea of society which is itself based on the notion that human and the majority of non-human beings, are all endowed with the capacity to act as autonomous agents. This is to say that the Batak universe is mapped by a complex network of social relations, which are not imagined as different manifestations of 'nature', but rather as different expressions of *kiaruwa*' which vary from one species to another, manifesting itself through different ways of thinking and behaving.

This does not, however, mean it is possible to draw a parallel between the Batak perspective of human-non-human interactions, and the recognisably ecocentric 'deep-ecologist' approach. In fact, the latter argue that not only do humans share a common physical substance with the other beings but, they also share a basic, intrinsic spiritual equality. Arne Naess (1989), the Norwegian philosopher most clearly associated with Deep Ecology, maintained that there are no ontological divides in the field of existence and that there is no dichotomy of values between humans and non-humans. Can this spiritual equality envisaged by deep ecologists be compared then, by any means, with the Batak notion of *kiaruwa*'?

We should be careful not to merge Batak perceptions of animals and plants with the position of radical conservationists who claim that all sentient beings are worthy of moral consideration in the same way that humans are. As I have proposed, in Batak thought the diversity of life forms has no value in itself outside the field of relationships in which humans and non-human agents are enmeshed. So, contrary to the Deep Ecology approach, Batak do not ascribe other living beings with intrinsic values but rather with relational values. Moreover, in Batak logic, the connectedness of all things is perceived as possible not by virtue of an abstract spirituality, but as a result of the individual beings' capacity 'to be a subject'. It follows that 'to be a subject' is the commonly shared condition, having its source in the *kiaruwa*'. A strong case can be made that biological egalitarianism endorsed by ecocentrists is not reflected in Batak ethics.

As we have seen, the way in which Batak shamans monitor the universe requires direct engagement with the world rather than 'distancing' and 'detachment' (cf. Ingold 1993), and thus the ability to travel across a landscape of social relationships where the common denominator of all living beings is not 'nature' (physical substance) but 'being subjects' (cf. Descola 1986, Viveiros De Castro 1998).

What is at stake becomes clearer when we think of the 'healing of the world' as the result of a 'complex agent'[6] at work. In fact, those taking part in this ceremony are all agents in their own right since they contribute by understanding and interpreting the shaman's gestures and words during his trance, and express statements and opinions on how he is performing his task at the *burungan*. However, no single statement or opinion can have control over the others, and interpretations and suggestions are always open to contestation. Remarkably, if the ritual is not successful and the storm persists, no one will impute the responsibility for failure to the shaman. In brief, Batak practices to 'heal the world' are not carried out by a single expert (the shaman), but involve a complex network of agents. This argument highlights a point of fundamental importance about the way in which agency is represented (Novellino 1999b), for it is not only the shaman, but other entities and community members who are also knowledgeable about local cosmology and myths, and about how these stipulate a set of social and economic principles. Re-establishing the socio-ecological balance thus also depends on the proper behaviour that community members and the guilty couple are expected to follow during the course and after the 'healing' ceremony. The incestuous couple is not a passive patient of the actions carried out by the shaman, rather they actively participate in the 'healing' performance (cf. Hobart 1993) offering chickens, plates and their own blood to the *tandayag* dragon.

A wider issue arises out of this discussion: local practices of 'healing the world' constitute local people as potential agents. This contrasts with how scientific knowledge, as often operationalised in 'global environmental management', represents conservation options (e.g., zoning) as technically superior, and the 'target beneficiaries' as passive recipients of such options (cf. Hobart 1993: 10).

Conclusion

The Western division between Nature and Culture takes an interesting guise when it is rehearsed under the framework of 'global' and 'local' debate. Nature, and all that it implies in modern conservationist discourse, is a universal resource, whose 'global' enjoyment and appreciation goes much behind the particular cultural context of those who have privileged access to such resources. 'Nature' is hence perceived as inherently public, objective rather than subjective, given rather than socially instituted. On a parallel level, it would appear that, in the Philippines, the enactment of environmental laws is not so much the product of a new political awareness but rather a cosmetic move to shift the terrain of discourse, so that national sovereignty becomes a form of 'caring for' rather then 'controlling' indigenous communities and their natural resources. In this way, state power has not been challenged; its vocabulary has simply been re-framed in environmentally friendly and politically accommodating terminology. This has legitimised the transformation of the indigenous territory into land categories systems, while maintaining the deceptive illusion that 'peoples' rights' are still retained in so-called 'tribal zones' (Novellino 1998, 1999a, 2000a, 2000b).

By and large, conservationist ideologies, both ecocentric and technocentric, seem to agree that 'nature' is the common denominator uniting humans with the rest of the living world in a ground of a common physical substance. There can be little doubt that any form of dualism of 'humans' and 'environment' makes little sense when the perceived reference point for both humans and non-humans is the capacity to act as subjects, rather than 'nature' as the common denominator. I would like to suggest that Batak are neither anthropocentrics nor ecocentrics. Indeed, in their metaphysics the scope of 'humanity' extends to a wide range of entities, including rice even, which are regarded as *taw*, or persons. Similarly, Batak are not ecocentrics, because they do not postulate humans and the rest of the cosmos as one undifferentiated set of mind and matter. More importantly, they do not regard themselves as elements of an ecological chain but rather as participants in a complex network of social relations taking place across overlapping ontological domains.

Regrettably, indigenous cosmological views are still being regarded by project planners and policy makers as merely folklore or legends. For Batak they are real: the migration of bees from the mythical *gunay gunay*, the revenge of the *tandayag* dragon, the shaman's voyage to the *burungan*, are facts rather than assumptions. Equally true is that, during the last few decades, Batak perceptions of the 'world' have widened up due to the increasing contacts with other societies. Their territory has dramatically shrunk because of land encroachment, creation of protected areas, and the limitations imposed on resource use by government ordinances and environmental laws.

Today, threats to the forest environment come unexpectedly from everywhere, and shamans feel that it is impossible for them to re-establish the cosmic balance when the sources of 'imbalances' are neither known nor detectable. People have come to realise that the world is much bigger than what the ancestors told them. Palawan is no longer the central layer of the universe but an island surrounded by many other 'islands' having different names: America, Europe, Japan, etc. As the people try to locate these worlds in their imagined map of the universe, they also begin to reflect on new questions: How far are such places from Palawan? What are the animals and plants living there? Do bees visit these places in the same way that they visit Palawan? And when the *burungan* is 'open' are such places also ravaged by storms and floods? These are questions that Batak have asked me, and to which I have had great difficulties in replying. Others, however, felt that such questions represented a long awaited opportunity and an ideal ground to enlighten the 'savages' on the issues of ecology, and on scientifically indisputable realities. To Batak, as of now, such realities remain painfully incommensurate.

Notes

1   The Batak are found scattered on a land area of about 240 square kilometres in the north-central portion of Palawan island (Philippines). Their population was estimated to comprise 380 individuals (Eder 1987), and continues to face demographic decline. They have a heterogeneous mode of food procurement, mainly centred on swidden cultivation and integrated with hunting, gathering and commercial collection of non-timber forest products. The case study at stake concerns the Batak community living within the territorial jurisdiction of Barangay Tanabag.

2   My analysis of Batak cosmology has benefited from the work of Howell (1984), and Viveiros De Castro (1998). Writing on Amerindian perspectivism, Viveiros De Castro has argued that 'the ability to adopt a point of view is undoubtedly a power of the soul, and non-humans are subjects in so far as they have (or are) spirit; but the differences between viewpoints ... lies not in the soul. Since the soul is formally identical in all species, it can only see the same things everywhere, the difference is given in the specificity of bodies' (1998: 478). A comparative study of perspectivism in both Amerindian and Asian cosmologies would highly advance anthropological understanding of indigenous metaphysics. For instance, Batak do not regard 'intangible essence' as the same in animals and in humans. Conversely, *kiaruwa'* is species-specific and endowed with a distinctive 'way of thinking'. On the other hand, human beings, masters of animals, benevolent *panya'en*, and even rice are all regarded as *taw* (persons), and thus capable of interacting on mutually intelligible grounds. Fieldwork carried out after this chapter will help to further elaborate Batak world views.

3   Palawan is the fifth-largest island in the Philippines, and has the highest percentage of forest cover in the archipelago, between 38 percent and 44 percent of the island surface (Serna 1990, Kummer 1992).

4   Different people may hold different opinions on whether plants have both *giinawan* (the breath) and *kiaruwa'*. As I was told by an informant 'all plants, as long as they are alive in the ground, they have the *kiaruwa'*. Without *kiaruwa'* they could not grow' (tape no. A, 9/3/1993). This same view has been confirmed by others. Further interviews have also revealed that the *kiaruwa'* of certain entities (*panya'en*) can dwell inside trees, bamboo and other plants. In relation to this, an elder from Tanabag, claims: 'If you cut a branch of a tree, and you see blood flowing out, it is a real person that you have cut!' (tape no. 1b, 11/8/1999). Furthermore, the shaman residing in Kalakuasan, maintains that in the area of Tagnipa there is a large log which is believed to be occupied by the *kiaruwa'* of a *tandayag* dragon (tape no. 28b, 21/5/1998). Overall, not everyone among the Batak is capable of providing detailed opinions on the most profound aspects of their worldviews, especially those dealing with the 'intangible components' of plants and other living beings.

5   In 1997 the Indigenous Peoples Rights Act (IPRA), also known as the Republic Act No. 8371, was enacted with the primary objective of recognising, protecting, and promoting the rights of indigenous cultural communities. Under the Estrada administration the IPRA law has finally sunk, and all pending applications for Certificates of Ancestral Domain Claim (CADC) have been frozen.

6   Hobart is referring to a 'complex agent' when 'decisions and responsibility for action involve more than one party in deliberation or action' (1990: 96).

11

ECOLOGISM AS AN IDIOM IN AMAZONIAN ANTHROPOLOGY

Stephen Nugent

Introduction

At first glance, the peoples of Brazilian Amazonia – Amerindians, north-eastern immigrants, historical peasantries and others – appear to be the subjects of several environmentally informed development projects mobilised by state and multilateral agencies such as the World Bank. However, thirty years after the construction of the Transamazon Highway, the unfolding of the Greater Carajás Project, and other massive efforts within the Plan for National Integration (PIN), it is still unclear if these peoples have had a role beyond that of being passive bystanders.

The Transamazon Highway, initially portrayed as a means for uniting people without land (northeasterners) and a land without people (Amazonia), was quickly revealed to be flawed (See Figure 3). Only a small percentage of the northeastern landless could be accommodated by a hastily arranged colonisation programme that was unsupported by either baseline research or adequate funding. Instead, the Transamazon became a brutal conduit between two versions of modern poverty, that of a declining plantation economy and that of the ecologically insecure independent frontier colonist relying on small, under-funded land holdings. There is no shortage of studies of the project, but the perspective of the colonists themselves has rarely been featured (Bunker 1978, 1985; Moran 1981, 1983). In these texts, the actors are portrayed as 'colonists', 'inhabitants', 'immigrants', or 'populations', as though their very coherence as motivated social actors was dissolved by successive modifications of 'the Plan'. One must mention that there has been much discussion of wise forest management, the lasting value of indigenous knowledge, and high-profile commercialisation of rainforest products by such companies as Body Shop International and Ben and Jerry's Ice-Cream. However, it is hardly obvious how this increased ecological discussion has brought about lasting results except, perhaps, as a peg for future research. Despite serious efforts to link social justice to environmental justice, actual, existing Amazonian societies continue to be portrayed as elements of a natural landscape – bit-actors in an epic natural Amazonia described as a 'green hell', 'the Last Frontier', 'the genetic warehouse', 'the lungs of the earth' or 'a carbon sink'.

The image of Amazonia as an epic natural space is the work of Victorian naturalists Richard Spruce, Henry Bates and Alfred Wallace. Through their long-term field research they provided systematic evidence of the biodiversity of the region, but also certain focal images – green hell, Conan Doyle's *Lost World* – which continue to captivate. The rise of a popular and scientific interest in environmentalism in the late twentieth century, and enormous research output from the region post-1970, placed Amazonia at the centre of a number of global discussions. The enormous biodiversity of Amazonia, for example, made it a prime case in the movement to not only protect endangered species, but to protect entire ecosystems. Speculations on the existence of special natural 'refugia', and on the kin-selection theories of E.O. Wilson and W. Hamilton, were developed in this environment (Forestra 1991). Accelerated rates of deforestation (over 12 percent of total forest cover in a period of twenty-five years) were cited as causes in the production of the greenhouse effect and were linked to large-scale hydrological changes, desertification, global climate change, and episodic phenomena such as El Niño. The net effect has been that the literary Amazonia of folklore, and the Amazonia of concerted, environmentally informed research, are two complementary and compelling images. The one is 'nature in the raw'. The other is nature filtered through increasingly sophisticated explanatory models that purport to achieve a bio-social synthesis. Thus we have not only a fetishistic image of Amazonia, but one further fetishised through scientific activity.

In this chapter I look at some of the ways in which anthropology in Amazonia has been reshaped by the new environmentalist agenda and argue that the supposed strength of the new synthesis of anthropological investigation supported by ecological sciences might be less promising that it appears. I will show why it is not surprising that actual existing Amazonian peoples still appear in scholarly literature as an out-of-focus or residual part of the research picture, and in popular literature as stereotypes such as befeathered Indians, bedraggled gold-miners, and frontier families in forest clearings.

Ecologising Amazonia

One consequence of the profusion of research literature[1] on Brazilian Amazonia produced over the past thirty years is that it is difficult to draw neat lines among the disciplines for which Amazonia holds special interest. Even a field as small as anthropology lacks precise definition. A collection like Anderson's *Alternatives to Deforestation* (1990), for example, offers numerous connections with anthropological research. The anthropological interest, however, typically remains confined to the idiom of human/cultural ecology (cf. Moran 1993). The state of the anthropological art, as depicted in three decennial overviews (Jackson 1975, Overing 1981, Viveiros de Castro 1996), is one in which there is a plurality of focal concerns, no one of which can claim dominance. This is not a condition of disarray, however, as much as it is a condition of the changed status of Amazonia in the international sphere, a sphere in which globalist environmentalist discourse has come to occupy a strategic position.

While it is often remarked that Amazonianist anthropology is only an amalgamation of case studies, there is a common cultural-ecological idiom according to which broad features of the social landscape – in pre-history and history – are explained by reference to the rigidities of the environment. In a succinct statement of this position, Meggers, in *Amazonia: Man and Culture in a Counterfeit Paradise* (1971) argues bluntly that the humid neo-tropics have insufficient carrying capacity for significant sociocultural development. Several factors are generally cited to support this idea. The first is the ethnographic record of the immediate post-Second World War period, compiled in Steward's *Handbook of South American Indians* (1949). A massive amalgamation of case studies, the *Handbook* succeeds in presenting Amazonia as a mosaic of delimited tribal societies whose overwhelming common feature is the natural setting (cf. Saloman and Schwartz 1999). The second is the lack of archaeological work in Brazilian Amazonia, hence relative ignorance about the degree to which contemporary Indian societies are representative of pre-history and pre-capitalist social formations (see Roosevelt 1994). The third is the prominence of cultural-materialist arguments of which the best known is the tendentious 'protein deficiency' variant (Gross 1975, Harris 1979). By this argument, the ceiling on hunter-gatherer cultural development in Amazonia reflects intrinsic shortcomings of the neotropical forest's carrying capacity.

While this 'counterfeit paradise' argument has been seriously challenged by Lathrap (1977), Beckerman and Kiltie (1980), Roosevelt (1991), and Balée (1998b), the idea that ecological constraint must be an overwhelming feature of social scientific explanation remains a central plank of the Amazonianist problematic. Even those anthropologists who do not directly engage the cultural–ecological idiom are perforce drawn into the fray. On one end of the spectrum we find Lévi-Strauss's Rousseauesque Nambikwara 'noble savages' and on the other, Chagnon's Hobbesian Yanomami 'fierce people' standing counterpoised as icons (Chagnon 1968, Lévi-Strauss 1971).[2]

The inclusion of studies of non-indigenous Amazonian societies (peasant, *caboclo*, frontier) as well as archaeological work and historical reconsideration has altered the research context. Although anthropology has moved away from a particularly rigid form of environmental determinism, the prevailing idiom in contemporary Amazonian studies, shaped by rising global environmentalist discourse, is a more sophisticated, but hardly less rigid, version of the same model. I shall refer to it as neo-environmental determinism, and later, as bio/sociodiversity. Thus, on one hand, anthropological theory in Amazonia has been emboldened by the new environmentalist turn (cf. Balée 1998b) attentive to a culture/nature dialectic best captured by the notion of anthropogenicism. On the other hand, it has been overwhelmed by an environmentalist discourse whose adherents are not simply researchers in cognate academic fields, but global policy-makers of considerable institutional power. Hence, the 'fate of the forest' is distributed amongst a diverse number of interest groups and academic fields. They are nominally united by an interest in the relationship between environmental conservation and social justice, but their material interests are actually quite divorced. The Tropical Forest Action Plan, promoted by the World Bank, and the biological reserves, promoted by the Rubber Tappers' Union, may be thematically affinal, but in all other respects are contradictory. Agenda-setting in Amazonia has primarily responded to the demands of two sets of interests: those representing the authoritarian regime of Brazil (1964–1985) and successor governments bearing the costs of two decades of clientelist restructuring. The global environmentalist lobby must embrace extremely diverse elements, ranging from the World Bank to local non-governmental organisations (NGOs), through the strategy of appealing to a common, green good.

From the local and the local to the local and the global

With cultural ecology established as a key framework for Amazonian anthropology, it was conventional to depict Amazonian societies as standing in a basically hostile relationship to nature. The possibilities for sociocultural development were taken to be highly constrained by intrinsic features of the neo-humid tropical forest ecosystem. Carneiro (1961) and Meggers (1971), for example, both make the case (albeit in different ways) that there is no purely social plane upon which anthropological investigation can proceed in Amazonia given that social agency is only tolerated so far before the environmental closes in. This is hardly a feature exclusive to work in Amazonia, but Amazonian ethnography has tended to produce or, often enough, leave implicit, fairly bald versions of this argument (by comparison, say, with Roy Rappaport's (1968) elaborate analysis of the dynamic interplay of social and ecological systems in Papua New Guinea). One consequence of this has been the production of fine-grained, pretty much self-contained ethnographic case studies against the background of an assumed, fairly uniform natural system. There have been studies exceeding these highly localised limits (Meggers 1971, Maybury-Lewis 1979, Lévi-Strauss 1973-81), but in general, the 'local' in Amazonian anthropology has infrequently had a 'global' counterpart until the relatively recent intervention of neo-environmental determinism. Neo-environmental determinism has produced a great increase in knowledge in lots of fields. Its impact on Amazonian societies has been substantially less marked despite the fact that environmental awareness is symbolically linked with social justice. In this regard neo-environmentalism is as much a theoretically sophisticated consolidation of environmental determinism as it is the basis for a critical, ecologically aware social science. What is crucially different between the two forms of environmental determinism is the fact that neo-environmentalism is globalist in perspective and driven less by scientific ambitions than by political ones. Thus, the environmental–ecological idiom which appears to unify, or at least provide a common referent for, contemporary research in Amazonia is perhaps not the uniformly progressive front so widely depicted. It is also a potentially passive acquiescence in favour of macro-social engineering, as testified by the Tropical Forest Action Plan or Brazil's *Avança Brasil* programme of accelerated resource extraction.

What Stephen Corry has aptly described as the 'rainforest harvest' (1993) makes a specious claim to virtue. Whether we consider debt for nature swaps, the promotion of biodiversity in the name of privileged access to tropical preciosities, or the expansion of non-value-added production in the region, all represent standard forms of exploitation of the region. Today they exploit under a mantle of heightened ecological awareness which is more style than substance, and ultimately does not deviate from a conception of Amazonia as the paragon of 'tropical nastiness' (Blaut 1994) and not as a socio-historical landscape. A major consequence of the high level of noise generated by environmentally informed development is that key political and scientific questions in Amazonia have been reduced to technical questions of management and administration (for example, the Brazilian government's rationalisation of *Avança Brasil*).

Two examples can be used to prove this point. The initiatives of the Rubber Tappers' Union in Acre to establish extractivist reserves where local knowledge of the environment is used to harvest local products bring in a modest income to peasants. This model garnered widespread support among scholars and activists who had studied first-hand the relationship between durable commercial exploitation of non-timber forest products and effective systems of land tenure. This model has instead been supplanted by that of national parks. The parks model is not environmentally uninformed, but does not represent the needs of actual forest-dwellers, but instead those within a planning apparatus. The planners see Amazonia as a gross source of export potential, not a set of coherent, actual-existing social systems. Under the neo-environmentalist agenda the direct exploiters of Amazonian resources – a peasantry whose transactions are often not market-mediated – are treated not as social agents engaged in the pursuit of regular livelihoods, but as poacher-appropriators. They are seen as actors who, distanced as they are from formal markets, are unlicensed scavengers. They appear as disreputable human add-ons to a vast resource still awaiting effective commercial administration.

Naïve space, non-standard antecedents

One of the assumptions underlying the hegemony of an ecological reading of contemporary Amazonia is that it is a naïve space. It is a place relatively unmarked by class structure or the hidden and overt injuries of modernity. This is a conceit which is accommodated with ease. The fact that there is an opera house in Manaus, for example, is not taken as a prompt to consider the significance of a hundred-year monopoly tenure of regional rubber extraction linked with industrial development in North America and Europe. Instead it is described as a typical Amazonian absurdity/anomaly. Here we can see a continuity in idiom with the anteaters and sloths dismissingly described by the naturalist Buffon as one maladaptation away from extinction.

The alleged naïvety of the social space does not survive even casual examination, but still endures as received wisdom. Several examples show the way that history is filtered by contemporary eco-driven discourse in Amazonia. The first of these is the rubber industry itself. It is often depicted as a 'boom'. In fact it was an enclave economy of great durability which integrated – via a commercial system dictated by the production demands of extractivism – Amazonia in a resolute if not permanent sense. Nor was this integration merely confined to the sphere of traders, rubber-tappers and tax-collectors, but was informed and shaped by imperial scientific and commercial entities such as Kew Gardens (see Brockway 1979). Another example, closely related to the structure of the rubber industry was the emergence of a pattern of urbanisation (Browder and Godfrey 1997) which sits uneasily with widely held conceits about the Amazonia of canoes, buttress roots and electric eels. Mitschein et al. (1989), in *Urbanizaçao Selvagem e Proletarização Passiva*, give us an ethnography of urban lifeways under the forest canopy which are quite devoid of protection. A third example, although superordinate, is the fact that Amazonia has been highly integrated within the world system for some centuries. The late-twentieth-century depiction of this 'green hell' as a novel site of contestation through 'sustainable development', 'empowerment' and 'environmental justice' simply does not adequately represent the social field.

Notably absent from both historical accounts and contemporary accounts of the 'conquest of the frontier' are the kinds of non-Amerindian societies which straddle the traditional/modern line. For, like so many of the examples elsewhere in this volume, they are people who cannot be contained within environmentally determinist or neo-environmentally determinist models – except with the greatest difficulty and distortion of the historical and ethnographic record. They are neither indigenes whose social character merely reflects, as some would have it, the rigidities of the natural system, nor are they recent frontier colonists whose precarious situation could be attributed to their failure to adapt to those same rigidities.

With regard to the historical accounts, Amazonia's legacy as a colonial region does have some unusual elements. The first is the ambiguity of its status as a colonial region. The area largely fell within the territory of the Spanish Empire (Treaty of Tordesillas, 1494), but was effectively occupied by the Portuguese, themselves representative of two empires, the Portuguese and the Brazilian, and later (1889) the Republic. The Amerindian population was almost entirely expunged during the course of colonisation. Those peoples who survived into the twentieth century were those who fled to the remote interior. Hence, in crucial respects, Amazonia was modernised at a very early stage. The society versus nature equation applied to indigenous peoples can hardly be an adequate generalisation for Amazonia at large.

Further, the colonial legacy of Amazonia is not typical of that of other regions of Brazil such as the slave-based plantation economies of the northeast or the agrarian, extractive and industrial economies of the centre-south. Largely disregarded by the State following the collapse of the rubber industry (c. 1912), Amazonia fell into what has been commonly diagnosed as a condition of 'economic stagnation'. It was briefly revalued as a source of *Hevea brasiliensis* (rubber trees) during the Second World War, but was not treated as an object of 'national integration' until Kubitschek's modernisation plan was instituted post-Second World War. The main point here is that colonial Amazonia has been far from the centre of national attention for most of its history and its administration has been largely by default rather than in the service of a predatory central power.

With regard to the historical peasantries of Amazonia there is a paucity of literature. Almost entirely ignored by anthropological researchers, *caboclos*, *ribeirinhos* and *seringueiros*[3] existed without benefit of much official or scientific attention. They were people who had the misfortune to be neither clearly nationals nor Amerindians.[4] The assault on Amazonia represented by Transamazonian Highway and related projects took little heed of the fact that Amazonia, far from being unoccupied and 'in need' of conquest ('installing the people without land in the land without people') was a complex colonial construct. In crucial respects, from the national point of view, Amazonians – Amerindian and neo-Amazonian – were non-peoples. The former were not people because they were and are wards of the State (Ramos 1998). The latter were not people because they were ersatz humans, *mestiços*, of the sort that so exercised Euclides da Cunha and Gobineaux (Schwarzc 1999). While it would be foolish to offer an Edenic counter-characterisation of Amazonian peasant life in mid-century, it is still the case that once the rubber boom spotlight was removed from Amazonia, social life did not go into suspended animation until given the Transamazonican kiss of life. Yet the narcoleptic caricature still prevails.

Nossa Natureza

Nossa Natureza was a policy initiative introduced by the administration of former President Jose Sarney, the first civilian President (although unelected) to take office following the withdrawal of the *golpista* generals from power (1985). *Nossa Natureza* (Our Nature) was regarded with scepticism and cynicism by many observers. It was clear that this homeopathic-strength palliative was a poorly thought out and superficial public relations exercise introduced by those responsible for allowing and encouraging environmental destruction on an unprecedented scale. One indicator of the scale of the assault is that over a twenty-five-year period of 'Amazonian modernisation' more than twelve percent of the forest cover was removed. This can be compared to the active preservation of the forest during the pre-historical occupation of Amazonia by Amerindians whose population approached that of contemporary Amazonia.

Dismissible though the *Nossa Natureza* initiative was as a substantive contribution, it exemplified a long-brewing change: Amazonia was not just a region in which the environmentalist/ecological imperative was pronounced. It was a key environmental region whose stewardship had implications for a global constituency as well as regional and national ones. In a series of authoritative publications commencing with Goodland and Irwin (1975), Amazonia's position in the world was recalibrated. Its new status impinged on long-term, evolutionary potential (biodiversity), meteorology (global warming, El Niño) and a host of macro-level issues. Implicated both in the assault on Amazonia, and critique of that assault, was an unprecedented level of research activity in the social and natural sciences, much of which operated – if only in general terms – within an ecological idiom. Even those researchers whose objects of analysis were only tangentially ecologically relevant could not but acknowledge at some level the ecological implications of their work.

It has been an assumption of much ecologically driven work in Amazonia that increased understanding of the ecosystem would illuminate social questions. Indeed this would seem to be the misconception behind the notion of 'sustainable development'. Here it is assumed that the better understanding of the limits of eco/biological systems allows accurate calculation of the demands social systems can place on nature. While not dismissing the laudable, if unachievable, goal of sustainable development, it does seem naïve and unrealistic to presume that eco-development is not entropic. It is no coincidence that the strongest appeals for policies of sustainable development come from the institutions and interest groups in the core, historically the major beneficiaries of development.

The pertinence of this to contemporary work in Amazonia is the following: existing models of sustainable development are consistently – but not always – disregarded in favour of hypothetical models of development informed by recognition of the ecological dimension (i.e., Pangloss with a gold card and a spread of ethical investments). The historical (and archaeological) record of accommodation to the constraints of green hell is bad, old news, while a quick-freeze method of *cupuaçu*[5] preservation is good news. The former is dull, complicated, and relatively unremunerative in the short-term. The latter is novel, possibly profitable, and maintains the traditional relationship (what some might call 'actual existing sustainable development') between the periphery and the core: the export of non-value-added primary materials. The engagement of social researchers (among them, anthropologists) in eco-driven reconceptualisations of the Amazonian social landscape is not, I hasten to add, to be derided. The outstanding question is whether they will achieve ecological fit (environmental justice and social justice) at the local level or merely reproduce traditional relations of subordination and dependency. The new lexicon of empowerment, grassroots, partnership, stakeholder, and so on, reflects the ambitions of the former, but semantic repackaging hardly effaces the significance of events such as occurred at El Dorado in 1996. Here members of the Landless Peoples Movement (MST: *Movimento Sem Terra*) were attacked and nineteen were killed. The articulation of such issues reveals the frailty of the eco-logic/socio-logic equation. Yet there is no pause in the modernisation programme. Thirty years after the joint birth of a programme of environmentally uninformed modernisation on the one hand and environmentally informed critique of that programme on the other, the former is only modestly mollified by the latter. Modernisation continues to employ the markers of ecologism and environmental awareness with success, and these are now integral to the hyper-real Amazonian social landscape.

One example of the partial rediscovery of the non-Amerindian social landscape has been the recuperation of the humblest of Amazonians. The peasant, once caricatured as *o caboclo indolente* (lazy half-breed), is now occasionally hailed as wise forest manager (along with his Indian counterpart). In part, this recuperation follows recognition that peasants who fish, cultivate swiddens, gather palm crops, and so on, actually maintain motivated rural livelihoods rather than subsist as virtual tropical scavengers. The success of low-level commodity producers who are able to accommodate production constraints placed on them by the annual flooding of the river-plains, high species diversity and low species density, remote markets, and many other limiting factors, has been gradually recognised. In addition, their successful manipulation of biological systems bespeaks a sophisticated understanding of complex social and biological domains rather more than the crude characterisation of 'indigenous and peasant knowledge'. What is less obvious, perhaps, is the fact that these now more visible systems have histories (cf. Adams, this volume). They do not just represent the outcomes of wily peasants quickly reacting to emergent commercial possibilities, but a highly structured peasant economy and – crucially from the perspective of environmental degradation – a sustainable economy.

A galvanising publication of the late 1980s was Peters et al.'s article in *Nature* (1989) on sustainable development in which they produced figures in support of the argument that the informal extraction of non-timber forest products (NTFPs) produced – over time – better incomes than could be got from using the same forest for either cattle raising or clear cutting. At the time, one line of criticism was that the claim was unrealistic given the politics of land tenure, uncertainty of markets, lack of infrastructure and a host of other elements that characterise many peasants' existence. What the article depicted better than much 'sustainable development' literature, however, was the fact that environmentally approved reckoning and calculation could not be divorced from the very real constraints on peasants' economic and political mobility. Producers of NTFPs are not voluntarily opting for palm fruit extraction over riding the range in a sports utility vehicle. Their social position licenses them to narrow and typically marginal domains, just as former rubber-tappers in the Brazilian Amazon did not become *caboclos* exploiting a variety of eco-niches out of some atavistic compulsion to keep a hand in forest work. A corollary is that efforts to make generalisations about the ecological appropriateness of peasant livelihoods based on a snap-shot – an almost compulsive feature of eco-dominated reasoning in Amazonia – run the risk of attenuating the range of factors relevant to the analysis of apparently stable systems. The issue here concerns not so much sustainability writ large (which is to say: in global terms, can economic growth be sustained without permanently subverting the long-term potential of the natural resource base?) as it concerns sustainability writ small (can precarious, poorly documented/understood economic sub-systems remain viable and still conserve their natural resource base?)

Amazonian social systems have traditionally been depicted as systems acquiescent to the rigid demands of the neo-humid tropical biosphere, hence the idea that there are not many Indians because the carrying capacity of the forest is so limited. Those new social systems that appear to contravene the limits of the biosphere (those which entail extensive deforestation for cattle pasture, open-pit mining or the predation of juvenile fish, etc.) are depicted as requiring management intervention to ensure that short-term commercial gain does not undermine the prospects for long-term exploitation. The historical record, however, complicates this neat picture of high contrast, for there are a number of intermediate possibilities.

First, work on the island of Combu in the Amazon estuary, where significantly higher than average incomes are achieved exclusively through the exploitation of *açaí* fruit, has attracted much interest, not least because of what appears to be a stable system of intensive palm forest management (Anderson 1990) (See Figure 3). The occupants of Combu have entertained a number of ecological investigations into diverse forms of resource management. The viability of *açaí* production on Combu, though, is not exhausted by a consideration of the interplay of ecological and social factors at the immediate, local-system level. The main factor is the burgeoning demand for *açaí* in the metropolis of Belém. This market demand reflects structural shifts in the regional economy and has little direct bearing on the ecology of Combu. Combu's viability as an enclave of wise forest management is shaped by a larger dynamic, yet one to which the residents of Combu have no direct access – or even necessarily knowledge.

A parallel example is provided by the once flourishing *cachaça*-producing industry of the estuary of which Combu forms a part. *Cachaça* – a major up-river trade item – was once extensively produced in the estuary in some one hundred and fifty distilleries whose energy was derived from tidal power. These distilleries were serviced by small-scale local producers for whom sugar-cane was one of a number of petty commodities. With the growth of the highway network, local production has succumbed to cheap imports from the south, and what was once an integral feature of the regional economy has become vestigial.

The further important factor is the prior livelihoods of contemporary Combu *açaí* producers. Most of the local domestic groups previously occupied terra firma swiddens in an adjacent, but ecologically quite different locale. Put another way, the success of Combu represents a fortunate, opportunistic shift that has less to do with the complex interplay of ecology and society in the estuary than it has to do with the valorisation of upland forest following the rise of agricultural demand in the periphery of Belém. On one hand, Combu looks like a nice, stable little system. On the other hand, it looks like a phase in which extractivism overshadows other rural livelihoods whose presence in the repertoire of production possibilities is suppressed for reasons which have little to do with the skills, desires, motivations or ambitions of the actors themselves. This is a form of sustainability, but not one that can be amplified into a general model sufficient to attract the interest of pan-Amazonian planners, but that shortcoming is formal, not substantive.

Another illustrative example is Tome-Açu (Figure 3). This is a Japanese colony celebrated for the introduction of black pepper cultivation in Amazonia. The economic success of the colonists is rarely attributed to the soundness of the relationship between ecosystem and social system. In fact, Tome-Açu is often portrayed as an anomaly: agriculturalists are not supposed to be able to sustain permanent cultivations for such a long period. Subler and Uhl (1990) note that what distinguishes Tome-Açu from the surrounding communities of non-Japanese Amazonian peasants is that in Tome-Açu cultivators have title to their land and a cooperative infrastructure. In other – and ostensibly more relevant – areas of expertise such as detailed knowledge of flora, Japanese colonists would appear to contradict simplistic assumptions about the merging of eco-logic and socio-logic. The sustainability of Tome-Açu is uncomfortable for both 'wise forest management' proponents and 'top-down managerial' experts, for it is inexplicable by recourse to either indigenous knowledge or modernisationist protocols.

A third illustration can be drawn from a recent piece concerning lowland Bolivian and Amazonian estuary agro-extractivists, in which Cleuren and Henkemans (2000) argue that successful peasant systems depend on their being 'intermediate'. Their meaning is that they represent an accommodation to economic, ecological and cultural constraints that are decidedly unstable. In other words, the rationality of peasant production is not wise management of the forest, but informed resistance to dependency on any one forest/river livelihood. The relative success of the estuary agro-extractivists seems to depend on sensitivity to the historical instability of product–market relations.

None of these examples is earthshaking, yet neither does any really fit the kind of derivative model of ecological anthropology which aspires to the satisfying incorporation of ecological analysis within anthropological analysis. In these examples there is no enhancement of anthropological explanation through greater sensitivity to systemic constraints which lie beyond the logic of the social system (in the biosphere). Yet a vocal case made in Amazonia (and elsewhere) today is for the inseparability of eco-logic and socio-logic summarised in the concept of sustainable development. While that is perhaps an admirable, but conceptually fraught goal, it also leads to a dangerous mystification, namely that the available models of ecology and anthropology are compatible just because one wishes them so.

Within the context of eco-discourse cited as evidence of the positive impact of an anthropology better informed by environmentalist concerns, there are two noteworthy developments. One of these is the growth of the NGO sector, especially those elements which function in the capacity of a kind of reflexive, applied anthropology. The second is the apparent political appeal of ecologism/environmentalism as an alternative to outdated platforms for social change. Chico Mendes – late president of the Rubber Tappers' Union – was portrayed by allies in the United States, for example, as a green warrior (which he was) rather than a trade union organiser (which he also was) on the grounds that the latter characterisation would lead him to fall victim to political marginalisation. They were probably right, but there could be no clearer indication of the importance of what is now seen as a rather old-fashioned kind of political calculation than the fact that Mendes died as a result of an old-fashioned bullet, not as a result of confusion over the spin factor.

Ecological agents, ecological structure

In the introduction to a recent volume, *Amazonia at the Crossroads: The Challenge of Sustainable Development*, Tony Hall writes that '...in the Amazonian context, sustainable development is a matter of conserving both biological diversity (biodiversity) and social diversity (sociodiversity) as two complementary sides of the same environmental coin' (2000: 2). This kind of formulation represents an ecologistic view increasingly expressed in recent years, not least in discourses of the globalisation of risk. It is also prominent in more technically focused arguments. According to Kidder and Balée, for example, in a programmatic account of historical ecology:

> The hallmarks of this emerging scientific transformation lie in two critical distinctions with regard to the normative scientific worldview: (1) no *a priori* separation between humans and nature can be empirically discerned, and (2) in regard to how we study the world, the distinction between natural and social sciences is tautological. The central theme that integrates the emergent science is a growing recognition that history and contingency are crucial concepts that need to be appreciated, and even embraced, if we are to enter into an effective scholarly dialogue within and among existing academic disciplines and comprehend the holism inherent in global ecology (1998: 405).

Laying aside, for the moment, scepticism with regard to the claim of obligatory holism (does nature really have a 'history', does nature transform society in the same way that society transforms nature?), what is striking about the claim is the implication it bears for the time-worn heuristic *agency versus structure*. This has profound consequences for the primary social actors within imperilled natural landscapes that have served as the focal examples for so much ecologically driven analysis, such as the peoples of Amazonia.

One of the major corollaries of formulations that aim to synthesise ecological and sociological perspectives in the manner prescribed above, is the establishment of a horizontal plane of analysis. This is not to discount continued and critical attention to hierarchy and inequality, but the globalisation position within which the eco/socio synthesis takes place grants priority to a systems-holism analysis which overshadows so-called old-fashioned *cui bono* considerations.

Amazonian Amerindian and peasant producers are characterised in the literature as having two quite different kinds of agency. The former are regarded as encapsulated by social structures hardened to the narrow environmental constraints of the humid neo-tropics, classic anthropological objects whose agency is severely attenuated by *habitus*. The latter are rarely granted acknowledgement of any agency. They do not even have a culturally integral agency – which Amerindians, for all their other shortcomings, do have. They are portrayed as populations, inhabitants, peasants whose arduous subsistence existence precludes anything as exalted as culture (cf. Nygren this volume).

What agency has been granted Amazonian peasant producers in recent years is a peculiar form of agency. It is not something that inheres in their social existence but in their having incorporated in their sociodiversity an appreciation of biodiversity. Hence, for example, the derided *caboclo indolente* (lazy river peasant) of previous decades now has the cachet of 'wise forest manager'. The newly granted eco-agency is, like many other Amazonian fauna, a curious beast. In the first place it represents a relatively undifferentiated sociodiversity. The specifics of Portuguese occupation, the Pombaline reforms promoting miscegenation, the emergence of *quilombos*, the competing claims of French, Dutch, English and other European entrants, Japanese colonists, the Jewish community, Lebanese traders, Italians, United States Confederates, United States corporate initiatives (e.g., Ford, Ludwig), the migration of *nordestinos* – to mention only a few of the most obvious underlying features of sociodiversity – are reduced to an identikit complement to a normative biodiversity which is itself not all that well understood. Hence we have a notion of Amazonian agency that stands not in a dialectical relationship with structure, but an agency that is the by-product of a new kind of structure, one which is a derivation of the virtuous pairing of eco-structure and socio-structure.

The new formulation of eco-agency has a number of progressive features, not least of which is the bare acknowledgment that peasant social systems in Amazonia represent more than passive reactions to the oppressive constraints of the *hyleia*.[6] Instead it could just as accurately be depicted as the rationalisation of a previously disorganised and ad hoc system for the extraction of primary tropical goods: increasing intervention in local-level systems of extraction in order to ensure predictable output. Agency is not attributed to actors within such integrated systems, it is attributed to the system itself.

The overriding question is: who are the direct beneficiaries of this new eco-concern? A definitive answer to that question is not available. The declared sociodiversity of the bio/sociodiversity formulation has no real content other than aggregate measures of the continued flow of tropical extractivist products to core consumers. The kind of significance granted to the effects of higher eco-consciousness on Amazonian direct producers may be inferred from the fact that so few studies have been carried out on these low-ranked systems, and this should not be surprising: the actors in these systems lack agency. They are bit-part players in another production, that of ensuring the viability of the eco/sociodiversity prescription as it seeks to meet the needs of those for whom Amazonia is a threatened source of export values. Hall (2000: 112), for example, notes that:

> As a complement to more conventional command-and-control mechanisms, pro-ductive conservation [i.e., sustainable development] has significantly widened the range of environmental policy options for Amazonia, capitalising on the potential of local populations to assist in finding more sustainable alternatives for securing livelihoods while protecting natural resources.

The argument presented here is that attention to the ecological ramifi-cations of development in Amazonia, while often linked with attempts to mitigate the effects of unregulated and uncontrolled despoliation, repre-sents actual Amazonian producers (indolent or sagacious) not as having agency, but as mere elements within a system (eco/sociodiversity). There are undoubtedly advances represented in an attempt to inform sociological analysis with ecological analysis, but there are strongly retrogressive features as well. Outstanding among them is the confirmation of the irrelevance of Amazonian sociality. Instead of a landscape populated by legatees of a diverse collection of Amazonian peasant (and other) societies, we are pre-sented with a new global resource:

> Biodiversity is an essential resource for adapting agricultural systems to shifting ecological and socio-economic conditions ... local knowledge is an often over-looked resource for the better management and conservation of biological resources ... The Amazon floodplain, or varzea, whose rich alluvial soils repre-sent the last major agricultural frontier of the Americas, provides a propitious setting for examining the interplay between biodiversity, agricultural intensifica-tion, and indigenous knowledge (Smith 1999: 4).

The citations above do not reflect the rhetorical posturing of entities such as the World Bank/IBDF, International Monetary Fund – hardly cir-cumspect in advancing their superior claims for global management and responsibility – but rather, programmatic statements from respected, expe-rienced and knowledgeable Amazonianists. Yet eco-discourse betrays the interests of the very people it is promoting. The proposed amplified system of conservation, sustainability, fair trade, planning, etc. is fully economised and adheres to a neo-liberal rule book, and the projection is that rational peasant production – carefully mobilised – will dovetail neatly with rational bio/sociodiversity planning. Yet who are the local economic actors queuing up for full participation?

This is a difficult question. Most Amazonianists would agree that such actors are a heterogeneous lot. Most would also agree that they have rather specialised and frequently unstable relations with local, regional, national and international markets, that they have uncertain legal claims to the land and resources upon which they depend (and as a result are high-risk economic partners), that they lack a common political platform, that they are inassimilable under a superordinate heading more precise than 'Amazonian'. They have been included in the new eco/socio-synthesis yet they appear to have no valence other than that conferred by nominal membership in the community of sociodiversity.

The horizontal platform created by the biodiversity/sociodiversity synthesis (late-modern human ecology) succeeds in placing in the same frame environmental justice and social justice. That is not an inconsiderable adjustment of the playing field, but it is an adjustment which does not necessarily attend to the needs and interests of local actors/systems whose functionality is defined by their perceived neo-ecological fitness (biodiversity + sociodiversity). A sanctioned role as guarantors of steady, sustainable output of low-cost forest and riverine produce hardly ameliorates their subject status, and certainly stops far short of mobilising ecological argument in the name of eradication of underprivilege. The scientific enhancement of the anthropological project provided by ecologism is considerable, but it is an enhancement that comes at some cost. As neo-ecology is increasingly shaped by a powerful managerial ethos reflecting the institutional weight of self-identified (and actual) global interests, anthropologically informed ecologism increasingly becomes defined by technique rather than critique. It is ironic that widely voiced scepticism about metanarratives in the human and social sciences should be so closely tracked by the meganarrative of neo-ecologism.

Notes

1   The general literature is vast as well, and often takes its lead from both professional and popular science publications. Portrayals of the exoticism of Amazonia have a numbing consistency. In the first month of the twenty-first century one may read in a broadsheet of the city of Manaus, 'one of the most remote cities in the world, 1,000 miles up the Amazon', a 'wild place where anacondas slithered down the street'. Coverage in non-Amazonian Brazil is not all that different.

2   It would be hard to characterise the profusion of an exceptionally rich and varied post-1960 ethnographic corpus as preoccupied with cultural/human ecological precepts. If anything, the dominant tendency has been toward the exploration of local knowledge systems, symbolism, kinship, social organisation, ritual and myth, relatively unencumbered by the theoretical implications of ecological imperative.

3   *Caboclo* = *mestiço* (for more elaborate discussion, see Nugent 1993), *ribeirinho* = riverbank-dweller, *seringueiro* = rubber-tapper. A neo-Amazonian might be all three, or none.

4   See Y. & R. Murphy's *Women of the Forest* (1974) for a particularly withering view of *caboclo* sociality.

5   *Cupuaçu* is closely related to *cacao* (chocolate) and has at times been promoted as an exportable tropical preciosity, mainly to be used in ice cream manufacture.

6   See Nugent (1993), Smith (1999), and Harris (2000) for empirical discussion.

BIBLIOGRAPHY

Abin, R. 1998. 'Plantations: village development threatens the survival of indigenous Dayak communities in Sarawak'. *Indigenous Peoples* 4: 15–23.

Abu Jaber, K., F. Gharaibeh, S. Khawasmeh, and A. Hill. 1978. *The Bedouin of Jordan: a People in Transition*. Amman: Royal Scientific Society.

Adams, C. 1994. 'As florestas virgens manejadas'. *Bol. Mus. Para. Emílio Goeldi, sér. Antropologia* 10(1): 3–20.

Adams, C. 2000a. *Caiçaras na Mata Atlântica pesquisa científica versus planejamento e gestão ambiental*. São Paulo: Annablume.

Adams, C. 2000b. 'As Populações Caiçaras e o Mito do Bom Selvagem a necessidade de uma nova abordagem interdisciplinar'. *Revista de Antropologia* 43(1): 145–82.

Adams, C. 2000c. As Roças e o Manejo da Mata Atlântica pelos Caiçaras uma revisão. *Interciência* 25(3): 143–50.

Adams, J. and T. McShane. 1992. *The Myth of Wild Africa: Conservation Without Illusion*. New York: W.W. Norton and Co.

Africare. 1993. *Rural water supply maintenance in the Kunene region*. Windhoek.

Agarwal, B. 1997. 'Environmental action, gender equity and women's participation'. *Development and Change* 28(1): 1–44.

Almeida A.P. 1946. 'Da decadência do litoral paulista'. *Revista do Arquivo Municipal*, 12 (107): 35–57.

Al-Sammane, H. 1981. *Al- Birnamij al-Suri li-Tahsin al-Mara'i wa Tarbiyat al-Aghnam* (Syrian programme for the improvement of range and sheep production). Damascus: Ministry of Agriculture and Agrarian Reform.

Amosov, A. 1998. *Evenki: Ocherk o gosudarstvennom regulirovanii zaniatnost malochislennykh narodov Severa v Evenkiiskom avtonomnom okruge*. Krasnoiarsk: Bukva.

Anderson, A. (ed.) 1990. *Alternatives to Deforestation*. New York: Columbia.

Anderson, D., and R. Grove (eds) 1987. *Conservation in Africa: People, Policies and Practice*. Cambridge: Cambridge University Press.

Anderson, D. G. 1998. *[First] field report: the marketing of wild meat in Taimyr and Evenkiia*. Unpublished manuscript. Circumpolar Liaison Directorate, Department of Indian and Northern Affairs.

Anderson, D. G. 2000a. *Identity and Ecology in Arctic Siberia: The Number One Reindeer Brigade*. Oxford: Oxford University Press.

Anderson, D. G. 2000b. 'Surrogate currencies and the wild market in Central Siberia'. In P. Seabright (ed.) *The Vanishing Rouble: Barter Networks and Non-Monetary Transactions in post-Soviet Societies*. Cambridge: Cambridge University Press.

Anderson, D. G. (In press). 'Nationality and 'Aboriginal Rights' in post-Soviet Siberia'. In T. Irimoto and T. Yamada (eds) *Ethnicity and Identity in the North*. Sapporo: Hokkaido University Press.

Anderson, D. G. and Ikeya, K. (eds) 2001. *Parks, Property, and Power: Managing Hunting Practice and Identity within State Policy Regimes*. Senri Publications in Ethnology. Vol. 59. Osaka: National Museum of Ethnology.

Antlov, H. 1995. *Exemplary Centre, Administrative Periphery: Rural Leadership and the New Order in Java*. Richmond: Curzon Press

Appadurai, A. (ed.) 1986. *The Social Life of Things. Commodities in Cultural Perspective*, Cambridge, Cambridge University Press.

Ashford, S. and L. Halman 1994. Changing attitudes in the European Community. In C. Rootes and H. Davis (eds) *A New Europe? Social Change and Political Transformation*. London: UCL Press, pp. 72–86.

Avança Brasil. www.infraestructurabrazil.gov.br

Balée, W. 1987. 'Cultural forests of the Amazon'. *Garden* 11(6): 12–14, 32.

Balée, W. 1992. 'Indigenous history and Amazonian biodiversity'. In H.K. Steen and R.P. Tucker (eds) *Changing Tropical Forests: Historical Perspectives on Today's Challenges in Central & South America*. Durham: Forest History Society, pp. 185–197.

Balée, W. 1995. 'Historical ecology of Amazonia'. In L.E. Sponsel (ed.) *Indigenous People and the Future of Amazonia: An Ecological Anthropology of an Endangered World*. Tucson & London: University of Arizona Press, pp. 97–110.

Balée, W. 1998a. 'Historical ecology: premises and postulates'. In Balée, W. (ed.) *Advances in Historical Ecology*. New York: Columbia, pp. 13–29.

Balée, W. (ed.) 1998b. *Advances in Historical Ecology*. New York: Columbia University Press.

Barnard, P. (ed.) 1998. *Biological Diversity in Namibia – A Country Study*. Windhoek: Nambian National Biodiversity Task Force.

Barsh, R. 1990. 'Indigenous peoples, racism and the environment'. *Meanjin* 49(4): 723–31.

Bebbington, A. 1997. 'New states, new NGOs? Crises and transitions among rural development NGOs in the Andean region'. *World Development* 25(11): 1755–65.

Beck, U, A. Giddens, and S. Lash, 1994. *Reflexive Modernization: Politics, Tradition and Aesthetics in the Modern Social Order*. Cambridge: Polity Press.

Beckerman, S., and R.A. Kiltie, 1980. 'More on Amazon cultural ecology'. *Current Anthropology* 21(4): 540–46.

Beinart, W. 2000. 'African history and environmental history'. *African Affairs* (99)395: 269–302.

Bender, B. 1993. 'Landscapes – meaning and action'. In B. Bender (ed.) *Landscape: Politics and Perspectives*. Oxford: Berg.

Benkhe, R., I. Scoones, and C. Kerven (eds) 1991. *Redefining Range Ecology: Drylands Programmes*. London: IIED.

Benkhe, R., I. Scoones, and C. Kerven (eds) 1993. 'Range ecology at disequilibrium: new models of natural variability and pastoral adaptation in African savannas'. London: Overseas Development Institute.

Bernardes, S. and Bernardes, C. 1950. 'A pesca no litoral do Rio de Janeiro'. *Revista Brasileira de Geografia*, Rio de Janeiro no. 2, pp. 54–65.

Biodiversity Support Program. 1997. *Biodiversity Conservation Network Annual Report 1997: Getting down to business*. Washington DC: Biodiversity Support Program.

Black Hills Alliance. 1980. *Black Hills/Paha Sapa report 2* (August/September).

Blaut, J.M. 1994. *The Colonizer's Model of the World*. London: Guildford Press.

Bollig, M. 1998. 'Power and Trade in Precolonial and Early Colonial Northern Kaokoland', In Hayes, P., J. Silvester, M. Wallace, and W. Hartmann, (eds) *Namibia Under South African Rule: Mobility and Containment 1915–46*, Oxford, Windhoek and Athens, James Currey, Out of Africa, Ohio University Press, pp. 175–93

Bonner, C. 1993. *At the Hand of Man: Perils and Hope for Africa's Wildlife*. New York: Alfred A. Knopf.

Botelle, A. and R. Rohde, 1995.*Those who live on the land: A socio-economic baseline survey for land use planning in the Communal Areas of Eastern Otjozondjup*. Land Use Planning Series: Report 1. Windhoek.

Bourdieu, P. 1984 (1979). *Distinction: A Social Critique of the Judgement of Taste*, trans. Richard Nice, London, Routledge & Kegan Paul.

Bowler, P. J. 1993. *Biology and Social Thought: 1850–1914*. Berkeley: Office for History of Science and Technology University of California at Berkeley.

Brandon, K., 1998. 'Perils to Parks: The Social Context of Threats'. In K. Brandon, K. H. Redford, and S. E. Sanderson (eds) *Parks in Peril: People, Politics and Protected Areas*. Washington DC and Covelo, California: The Nature Conservancy/Island Press, pp. 415–39.

Brandon, K., K.H. Redford, and S.E. Sanderson. 1998. 'Introduction'. In K. Brandon, K.H. Redford and S.E. Sanderson (eds) *Parks in Peril: People, Politics and Protected Areas*. Washington DC: Island Press, pp. 1–23.

Braun, B. and N. Castree (eds) 1998. *Remaking Reality: Nature at the Millenium*. London and New York: Routledge.

Bridger S. and F. Pine, 1998. *Surviving post-socialism: Local Strategies and Regional Responses in Eastern Europe and the Former Soviet Union*. London: Routledge.

Brockway, L. 1979. *Science and Colonial Expansion*. New York: Academic Press.

Brosius, P. J. 1999. 'Analyses and interventions: Anthropological engagements with environmentalism'. *Current Anthropology* 40 (3): 277–309.

Brosius, P.J., A.L. Tsing, and C. Zerner, 1998. 'Representing communities: Histories and politics of community-based natural resource management'. *Society and Natural Resources* 11: 157–68.

Browder, J., and B. Godfrey, 1997. *Rainforest Cities*. New York: Columbia University Press.

Brown K.S., and G.G. Brown. 1992. 'Habitat alteration and species loss in Brazilian forests'. In T.C. Sayer (ed.) *Tropical Deforestation and Species Extinction*. London: Chapman and Hall, pp. 119–42.

Bryant, B., and P. Mohai, 1992. 'Summary'. In Bryant, B. and P. Mohai (eds) *Race and the Incidence of Environmental Hazards: A Time for Discourse*. Boulder: Westview Press, pp. 215–19.

Bryant, D., D. Nielsen, and L. Tangley, 1997. *The Last Frontier Forests: Ecosystems and Economies on the Edge*. Washington DC: World Resource Institute.

Bullard, R. 1990. *Dumping in Dixie: Race, Class and Environmental Quality.* Boulder: Westview Press.
Bunker, S. 1983. 'Policy implementation in an authoritarian state: A case from Brazil'. *Latin American Research Review*, 18(1): 33–58.
Bunker, S. 1985 *Underdeveloping the Amazon: Extraction, Unequal Exchange and the Failure of the Modern State.* Urbana and Chicago: University of Illinois Press.
Callihan, D. 1999. *Using tourism as a means to sustain community-based conservation: experience from Namibia.* Report for the LIFE (Living in a Finite Environment) project, Windhoek.
Campbell, C.A.R. 2001. 'Transits and tenures'. Unpublished MA thesis, Department of Anthropology, University of Alberta.
Campos, F.P. 1980. *A situação dos posseiros de Trindade no litoral sul fluminense uma visão histórica.* Ma. diss., Escola de Comunicações e Arte da Universidade de São Paulo.
Cândido, A. 1964. *Os parceiros do Rio Bonito. Estudos sobre o caipira paulista e a transformação dos seus meios de vida.* São Paulo: Duas Cidades.
Carneiro, R. 1961. 'Slash-and-burn cultivation among the Kuikuru and its implications for cultural development in the Amazon Basin'. In J. Wilbert (ed.) *The Evolution of Horticultural Systems in Native South America: Causes and Consequences.* Caracas, Venezuela: Antropologica Supplement No. 2 pp. 47–68.
Carrier, J. G. and J. McC. Heyman, 1997. 'Consumption and political economy', *Journal of the Royal Anthropological Institute*, (N.S.) 3: 355–73.
Carruthers, J. 1995. *The Kruger National Park: A Social and Political History.* Pietermaritzburg, University of Natal Press.
Chagnon, N. 1968. *Yanomamo: The Fierce People.* New York: Holt Rinehart and Winston.
Chambers, R., and P. Richards. 1995. Preface. In D. M. Warren, L. J. Slikkerveer, and D. Brokensha, (eds) *The Cultural Dimension of Development: Indigenous Knowledge Systems.* London: Intermediate Technology Publications, pp. xiii–xiv.
Chatty, D. 1986. *From Camel to Truck.* New York: Vantage Press.
Chatty, D. 1990. 'The current situation of the Bedouin in Syria, Jordan and Saudi Arabia and their prospects for the future'. In C. Salzman and J. Galaty (eds) *Nomads in a Changing World.* Naples: Istituto Universitario Orientale, Series Minor, pp. 123–37.
Chatty, D. 1995. *Hired Shepherds: The Marginalization and Impoverishment of Pastoralists in Jordan and Syria.* Amman: CARDNE.
Chaui, M. 2000. *Brasil. Mito Fundador e Sociedade Autoritária* (vol. 1). São Paulo: Editora Fundação Perseu Abramo.
Chavis, Benjamin F. (Jr). 1993. 'Foreword'. In Robert Bullard (ed.) *Confronting Environmental Racism: Voices from the Grassroots.* Boston: South End Press, pp. 3–5.
Churchill, W. and W. LaDuke, 1992. Native North America: the Political Economy of Radioactive Colonialism. In Annette Jaimes, M. (ed.) *The State of Native America: Genocide, Colonization, and Resistance.* Boston: South End Press, pp. 241–66.
Cleary, D. 1993. 'After the frontier: Problems with political economy in the modern Brazilian Amazon'. *Journal of Latin American Studies*, 25(2): 331–49.
Cleuren, H.M., and A.B. Henkemans, 2000. *The Resilience of Agro-Extractivist Systems of the Cambas and Caboclos of Today's Amazon Forest.* Http://org.nlh.no/etfrn/lofoten/cleurenpap.htm, (accessed August 2000).
Colchester, M. 1999. 'Sharing power: Dams, indigenous peoples and ethnic minorities'. *Indigenous Affairs*, 3/4: 4–54.
Conklin, B. 1997. 'Body paint, feathers, and VCRs: Aesthetics and authenticity in Amazonian activism'. *American Ethnologist*, (24): 711–37.
Corbridge, S., N. Thrift, and R. Martin (eds) 1994. *Money, Power and Space.* Oxford: Blackwell.
Corry, S. 1993. 'The rainforest harvest: Who reaps the benefit?' *The Ecologist*, 23(4): 148–53.
Cosgrove, D. 1993. 'Landscapes and myths, gods and humans'. In B. Bender (ed.) *Landscape Politics and Perspectives.* Oxford: Berg, pp. 281–306.
Costa, M.B.B. da, 1991. *Contribuição à Formulação de Plano Diretor e Programa de Ação para Agricultura e Extrativismo na Estação Ecológica de Juréia-Itatins.* São Paulo: Secretaria do Meio Ambiente.
Cronon, W. (ed.) 1995. *Uncommon Ground: Toward Reinventing Nature.* London: W.W. Norton & Company.
Cronon, W. 1996. 'The trouble with wilderness or, getting back to the wrong nature'. In W. Cronon (ed.) *Uncommon Ground: Rethinking the Human Place in Nature.* New York: W.W. Norton & Company, pp. 69–90.

Cronon, W., G. A. Miles, and J. Gitlin, (eds) 1992. *Under an Open Sky: Rethinking America's Western Past*. 1st ed. New York: W.W. Norton.

Crook, C. and R.A. Clapp. 1998. 'Is market-oriented forest conservation a contradiction in terms?' *Environmental Conservation*, 25(2): 131–45.

Cunha L.H.O. and M.D. Rougeulle, 1989. *Comunidades litorâneas e unidades de proteção ambiental convivência e conflitos; o caso de Guaraqueçaba (Paraná)*. São Paulo: PPCAUB – Universidade de São Paulo.

DANIDA. 1998. *Documento de Proyecto Manejo Sostenible de la Zona de Amortiguamiento, Nicaragua*. Managua: DANIDA (mimeogr.)

Davis, R. and M. Zannis, 1973. *The Genocide Machine in Canada*. Montreal: Black Rose.

DEA (Directorate of Environmental Affairs, Namibia) http://www.dea.met.gov.na/programmes/ cbnrm/ conservancies.htm (accessed April 2001).

Dean W. 1996. *With Broadax and Firebrand: The Destruction of the Brazilian Atlantic Forest*. Los Angeles: University of California Press.

Debaine, F. 2000. The Degradation of the Steppe Areas in Northwestern Syria: Perceptions and Reality. Paper presented at the Middle East Studies Association annual conference. November 2000, Orlando, Florida, United States.

Descola, P. 1986. *La Nature domestique: Symbolisme et praxis dans l'écologie des Achuar*, Paris: Maison des Sciences de l'Homme.

Descola, P. 1993. *Le Lances du Crepuscule: Relations Jivaros, Haute-Amazonie*. Paris: Plon.

Descola, P. and G. Pálsson, 1996a. 'Introduction'. In P. Descola and G. Pálsson (eds) *Nature and Society. Anthropological Perspectives*. London: Routledge, pp. 1–21.

Descola, P, and G. Pálsson (eds) 1996b. *Nature and Society: Anthropological Perspectives*. New York: Routledge.

Dewdney, R. 1996. *Policy Factors and Desertification: Analysis and Proposals*. Windhoek, NAPCOD.

Diamond, J.M. 1987. 'The environmentalist myth'. In *Nature* 324: 19–20.

Di Chiro, G. 1992. 'Defining environmental justice: Women's voices and grassroots politics'. *Socialist Review* 22(4): 93–130.

Didier, G. 1993. *Etude de la dynamique des systèmes de production sur la frontière agricole du département du Río San Juan, Nicaragua*. Département d'Agronomie, Université de Montpellier (unpublished).

Diegues A.C.S. 1983. *Pescadores, camponeses e trabalhadores do mar*. São Paulo: Ática.

Diegues A.C.S. 1988. *Diversidade biológica e culturas tradicionais litorâneas o caso das comunidades caiçaras*. São Paulo: NUPAUB – Universidade de São Paulo.

Diegues A.C.S. 1993. *Populações tradicionais em unidades de conservação o mito moderno da natureza intocada* (vol. 1). São Paulo: NUPAUB – Universidade de São Paulo.

Diegues A.C.S. and P. Nogara, 1994. *O nosso lugar virou parque estudo sócio-ambiental do Saco de Mamanguá*, Parati. Rio de Janeiro/ São Paulo: NUPAUB – Universidade de São Paulo.

Dobson, A. 1990. 'Thinking about ecologism'. In A. Dobson (ed.) *Green Political Thought*. London: Harper Collins Academic, pp. 13–37.

Dobson, R.B. 1994. 'Communism's legacy and Russian youth'. In J.R. Millar and S.L. Wolchik (eds) *The Social Legacy of Communism*. Cambridge MA: Cambridge University Press, pp. 229–52.

Douglas, M. and B. Isherwood, 1978. *The World of Goods: Towards an Anthropology of Consumption*, London, Penguin.

Dove, M.R. 1993. 'A revisionist view of tropical deforestation and development'. *Environmental Conservation*, 20(1): 17–24/56.

Draz, O. 1977. *Role of Range Management and Fodder Production*. Beirut: UNDP Regional Office for Western Asia.

Durbin, J., B.T.B. Jones, and M. Murphree, 1997. *Namibian Community-Based Natural Resource Management Programme: project evaluation 4–19 May 1997*. Unpublished report submitted to Integrated Rural Development and Nature Conservation (IRDNC) and World Wide Fund for Nature (WWF), Windhoek.

EarthAction. http://www.earthaction.org/en/archive/97-04-ForYam/alert.html (accessed October 1997).

Eckersley, R. 1992. *Environmentalism and Political Theory: Towards an Ecocentric Approach*. London: University College London Press.

Eder, J.F. 1987. *On the Road to Tribal Extinction: Depopulation, Deculturation, and Maladaption among the Batak of the Philippines*. Berkley: University of California Press.

Edwards, M. and D. Hulme, 1996. 'Too close for comfort? The impact of official aid on nongovermental organizations'. *World Development*, 24 (6): 961–73.

Ehrenfeld, D. 1988. 'Why put a value on biodiversity?' In E.O. Wilson (ed.) *Biodiversity*. Washington D. C.: National Academy Press, pp. 212–16.

Ellen, R.F. 1986. 'What Black Elk left unsaid: On the illusory images of green primitivism'. *Anthropology Today*, 2 (6): 8–12.

Ellen, R.F. 1993. 'Rhetoric, practice and incentive in the face of the changing times: a case study in Nuaulu attitudes to conservation and deforestation'. In K. Milton (ed.) *Environmentalism: The View from Anthropology*. London: Routledge, pp. 126–43.

Ellen, R and K. Fukui (eds) 1996. *Redefining Nature: Ecology, Culture and Domestication*. Oxford: Berg.

Ellis, S. 1994. 'Of elephants and men: Politics and nature conservation in South Africa'. *Journal of Southern African Studies*, (20)1: 53–69.

Escobar, A. 1995. *Encountering Development: The Making and Unmaking of the Third World*. Princeton: Princeton University Press.

Escobar, A. 1996. 'Constructing nature: Elements for a poststructural political ecology'. In Richard Peet and Michael Watts (eds), *Liberation Ecologies: Environment, Development, Social Movements*. London: Routledge, pp. 46–68.

Escobar, A. 1999. 'After nature: Steps to an antiessentialist political ecology'. *Current Anthropology*, 40(1): 1–30.

Fabian, M. 1983. *Time and the Other: How Anthropology Makes its Object*. New York: Columbia University Press.

Fairhead, J. and M. Leach, 1995. 'False forest history, complicit social analysis: Rethinking some West African environmental narratives'. *World Development*, 23(6): 1023–35.

Fairhead, J. and M. Leach, 1996. *Misreading the African landscape: Society and ecology in a forest-savanna mosaic*. Cambridge: Cambridge University Press.

FAO. 1965. Land policy in the Near East. Rome: FAO.

FAO. 1972a. Near East regional study: Animal husbandry production and health, fodder production and range management. Rome: FAO.

FAO. 1972b. FAO expert consultation on the settlement of nomads in Africa and the Near East. Cairo: FAO.

FAO. 1995. Rangeland rehabilitation and establishment of a wildlife reserve in Palmyra Badia (Al-Taliba). Document no. GCP/SYR/003. Rome: FAO.

Ferguson, J. 1990. *The Anti-Politics Machine: 'Development', Depoliticization, and Bureaucratic Power in Lesotho*. Cambridge: Cambridge University Press.

Filer, C. 1990. 'The Bougainville rebellion, the mining industry, and the process of social disintegration in Papua New Guinea'. *Canberra Anthropology*, 13 (1): 1–39.

Filer, C. 1991. 'Two shots in the dark: The first year of the task force on environmental planning in priority forest areas'. *Research in Melanesia*, 15(1): 1–48.

Fine B., and E. Leopold, 1993. *The World of Consumption*. London: Routledge.

Fisher, W.F. 1997. 'Doing good? The politics and antipolitics of NGO practices'. *Annual Review of Anthropology*, 26: 439–64.

Fondahl, G. 2000. *Boundaries and Identities: Reconceptualising Social Spaces in Berezoka Nasleg, Northeast Siberia*. Paper presented at the conference 'Postsocialisms in the Russian North'. Max Planck Institute for Social Anthropology, Halle (Saale) Germany.

Forbes, A. 1987 [1887]. *Unbeaten Tracks in the Islands of the Far East: Experiences of a Naturalist's Wife in the 1880s*. Singapore: Oxford University Press.

Forbes, H.O. 1989 [1885]. *A Naturalist's Wanderings in the Eastern Archipelago*. Singapore: Oxford University Press.

Foresta, R. 1991. *Amazon Conservation in the Age of Development: The Limits of Providence*. Gainesville: University of Florida Press.

França, A. 1954. *A Ilha de São Sebastião. Estudo de Geografia Humana*. São Paulo: FFCL – Universidade de São Paulo.

France. 1923–1938. *Minstère des Affaires Etrangères, Rapport sur la situation de la Syrie et du Liban soumis au conseil de la Societe des Nations 1923–1938*.

Franklin, S. C. Lury, and J. Stacey. 2000. *Global Nature, Global Culture*. London, Thousand Oaks, New Delhi: SAGE.

Freeman, Milton M.R., L. Bogoslovskaya, R. Caulfield, I. Egede, I. Krupnik, and M. Stevenson. 1998. *Inuit, Whaling, and Sustainability*. Walnut Creek: Altamira Press.

Friedman, J. (ed.) 1994. *Consumption and Identity*. Switzerland: Harwood Academic Publishers.

Fuller, B.B. Jnr, 1993. *Institutional Appropriation and Social Change Among Agropastoralists in Central Namibia 1916–1988*. Unpublished PhD thesis, Graduate School, Boston.

Fundação SOS Mata Atlântica. 1992. *Dossiê Mata Atlântica 1992.* São Paulo: Fundação SOS Mata Atlântica.

Furtado. C. 1968. *The Economic Growth of Brazil: A Survey from Colonial to Modern Times.* Berkeley and Los Angeles: University of California Press.

Gadgil, M. and R. Guha, 1994. Ecological conflicts and the environmental movement in India, *Development and Change*, 25(1): 101–36.

Gaiulskii, Artur I. 1999. *Doklad na s"ezd seliian i sel'khoztovaroproizvoditelei.* Unpublished conference paper. First Congress of Villagers and Rural Producers. Tura, Evenkiia April 15, 1999.

Garland, E. and R.J. Gordon. 1999. The authentic (in)authentic: Bushman anthro-tourism. *Visual Anthropology*, (12): 267–87.

Gedicks, Al. 1993. *The New Resource Wars: Native and Environmental Struggles Against Multinational Corporations.* Boston: South End Press.

Gedicks, Al. 1994. 'Racism and resource colonization'. *Capitalism, Nature, Socialism*, 5 (1): 55–76.

Gell, A. 1996. 'The language of the forest: Landscape and phonological iconism in Umeda'. In E. Hirsch and M. O'Hanlon (eds) *The Anthropology of Landscape.* Oxford: Clarendon Press, pp. 232–54.

George, S. 1998. 'Preface'. In M. Goldman (ed.) *Privatizing Nature: Political Struggles for the Global Commons.* London: Pluto Press.

Ghimire, K.B., and M.P. Pimbert. 1997. 'Social change and conservation: an overview of issues and concepts'. In K.B. Ghimire and M.P. Pimbert (eds) *Social Change and Conservation.* London: Earthscan, pp. 1–45.

Gibson, J.J. 1979. *The Ecological Approach to Visual Perception.* Boston: Houghton Mifflin.

Giddens, Anthony. 1991. *Modernity and Self-Identity: Self and Society in the Late Modern Age.* Cambridge: Polity Press.

Gill, S.D. 1987. *Mother Earth: An American story.* Chicago: University of Chicago Press.

Goodland, R. 1975. 'History of "ecology"'. *Science*, (188): 313.

Goodland, R., and H.S. Irwin. 1975. *The Amazon Jungle: From Green Hell to Red Desert?* New York: Elsevier.

Gordon, R.J. 1991. Vernacular Law and the Future of Human Rights in Namibia. NISER *Discussion Paper* 11. Namibian Institute for Socio-Economic Research, Windhoek

Gordon, R.J. 2000. 'The stat(u)s of Namibian anthropology: A review'. *Cimbebasia* 16: 1–23.

Gottlieb, R. 1993. 'Reconstructing environmentalism: complex movements, diverse roots'. *Environmental History Review*, 17(4): 1–19.

Government of South Africa 1964. Report of the commission of inquiry into South West African affairs. Pretoria, Government Printer.

Government of the Republic of Namibia 1991. *The Constitution.* Windhoek: Republic of Namibia.

Gray, P. 2000. *Chukotkan reindeer husbandry in the twentieth century: In the image of the Soviet economy.* Paper presented at the 12th Inuit Studies Conference, University of Aberdeen.

Grey Owl, M. 26 March 1998. *Intervention to the UN Commission on Human Rights*, Geneva, Switzerland. Wolf Creek, SD: Black Feather Private Collection.

Gross, D. 1975. 'Protein capture and cultural development in the Amazon Basin'. *American Anthropologist*, 77(3): 526–49.

Guha, Ramachandra. 1990. *The Unquiet Woods: Ecological Change and Peasant Resistance in the Himalayas.* Berkeley, CA: University of California Press.

Guha, Ramachandra, and J. Martínez-Allier, 1997. *Varieties of Environmentalism: Essays North and South.* London: Earthscan.

Gupta, A. 1998. *Postcolonial Developments: Agriculture in the Making of Modern India.* Durham, NC: Duke University Press.

Gupta, A., and J. Ferguson (eds) 1997. *Culture, Power, Place: Explorations in Critical Anthropology.* London: Duke University Press.

Haacke, W., E. Eiseb, and L. Namaseb, 1997. 'Internal and external relations of Khoekhoe dialects: A preliminary survey'. In W.H.G. Haacke and E.D. Elderkin (eds) *Namibian Languages: Reports and Papers*, Windhoek: Gamsberg Macmillan and Köln, pp. 125–209.

Haacke, W.H.G. and E. Eiseb, 1999. *Khoekhoegowab-English Glossary/Midi Saogub*, Windhoek: Gamsberg Macmillan and Köln.

Hadenius, A. and F. Uggla, 1996. 'Making civil society work, promoting democratic development: What can states and donors do?' *World Development*, 24 (10): 1621–39.

Hall, A. (ed.) 2000. 'Environment and development in Brazilian Amazonia'. In Hall (ed.), *Amazonia at the Crossroads.* London: ILAS, pp. 99–114.

Hann, C.M. (ed.) 1998. *Property Relations: Renewing the Anthropological Tradition.* Cambridge: Cambridge University Press.

Harris, Mark. 1998. 'What it means to be 'caboclo''. *Critique of Anthropology,* 18 (1): 83–95.

Harris, Mark. 2000. *Life on the Amazon.* Oxford: Oxford University Press.

Harris, Marvin. 1979. *Cultural Materialism.* New York: Random House.

Harrison, R. Pogue. 1992. *Forests: The Shadow of Civilization.* Chicago: University of Chicago Press.

Harvey, D. 1989. *The Condition of Postmodernity: An Enquiry into the Origins of Cultural Change.* Oxford: Basil Blackwell.

Harvey, D. 1993. 'The nature of environment: The dialectics of social and environmental change'. In R. Miliband and L. Panitch (eds) *Socialist Register.* London: Merlin, pp. 1–51.

Harvey, D. 1996. *Justice, Nature and the Geography of Difference.* Oxford: Blackwell Publishers.

Hayden, C. P. 1998. 'A Biodiversity Sampler for the Millennium. In S. Franklin and H. Ragoné (eds) *Reproducing Reproduction: Kinship, Power, and Technological Innovation.* Philadelphia PA: Pennsylvania U.P., pp. 173–206.

Higgs, E. 2000. 'Nature by design'. In E. Higgs, A. Light, and D. Strong (eds) *Technology and the Good Life?* Chicago: University of Chicago Press, pp. 195–212.

Hill, J.D. 1996. 'Introduction: Ethnogenesis in the Americas'. In J.D. Hill (ed.) *History, Power, and Identity.* Iowa: University of Iowa Press, pp. 1–19.

Hladik, C. M., A. Hladik, O. F. Linares, H. Pagezy, A. Semple, and M. Hadley, (eds), 1993. *Tropical Forests, People and Food: Biocultural Interactions and Applications to Development.* Man and the Biosphere series, Volume 13. UNESCO: Paris.

Hobart, M. 1990. 'The patience of plants: A note on agency in Bali'. *Review of Indonesian and Malaysian Affairs,* vol. 24(2): 90–135.

Hobart, M. 1993. 'Introduction: The growth of ignorance?' In M. Hobart (ed.) *An Anthropological Critique of Development.* London: Routledge, pp. 1–30.

Homewood, K.M. and W.A. Rodgers, 1987. 'Pastoralism, Conservation and the Overgrazing Controversy'. In D. Anderson and R. Grove (eds) *Conservation in Africa: People, Policies and Practice.* Cambridge: Cambridge University Press, pp. 111–28.

Howard, A. and J.M. Mageo (eds) 1996. *Spirits in Culture, History and Mind.* London: Routledge.

Howell, S. 1984. *Society and Cosmos: Chwong of Peninsular Malaysia.* Oxford: Oxford University Press.

Howell, S. 1996. 'Nature in culture or culture in nature: Chewong ideas of "humans" and other "species"'. In P. Descola and G. Pálsson (eds) *Nature and Society. Anthropological Perspectives.* London: Routledge, pp. 127–44.

Hyde, G.E. 1937. *Red Cloud's Folk: A History of the Oglala Sioux Indians.* Norman, OK: University of Oklahoma Press.

Hyde, G.E. 1956. *A Sioux Chronicle.* Norman, OK: University of Oklahoma Press.

ICTI 1993. *Potensi dan Peluang Pembangunan Pulau Yamdena, Maluku Tenggara, Propinsi Maluku,* Jakarta: ICTI.

ILO. 1964. *Technical meeting on problems of nomadism and sedentarisation.* Geneva: ILO.

Ingold, T. 1986. *The Appropriation of Nature: Essays on Human Ecology and Social Relations.* Manchester: Manchester University Press.

Ingold, T. 1993. 'Globes and spheres: The topology of environmentalism'. In K. Milton (ed.) *Environmentalism: The View from Anthropology.* London: Routledge, pp. 31–42.

Ingold, T. 2000. *The Perception of the Environment: Essays in Livelihood, Dwelling and Skill.* London: Routledge.

Inambao, C. 1998. Torra on Threshold of Brighter Future, *The Namibian,* 9 September, Windhoek.

Inhutani 1, http://www.inhutani1.co.id (accessed September 2000).

Institute of Petroleum. 1998. *Emerging markets for emissions trading: Opportunities from the Kyoto Protocol and the implications for business. Complete proceedings from an international conference held in London, 11–12 May 1998.* London: Institute of Petroleum.

IPARDES. 1989. *APA de Guaraqueçaba caracterização sócio-econômica dos pescadores artesanais e pequenos produtores rurais.* Curitiba: IPARDES.

IRDNC n.d. *Information Document* Windhoek: IRDNC.

IRENA 1992. *Plan de acción forestal: Documento base.* Managua: IRENA.

IUCN/UNEP/WWF. 1980. *World Conservation Strategy: living resource conservation for sustainable development.* Gland, IUCN.

IUCN Bulletin (no. 3, 1993)

Jackson, J. 1975. 'Recent ethnography of indigenous Northern Lowland South America'. *Annual Review of Anthropology,* 4: 307–340.

Jacobs, A. 1975. 'Maasai pastoralism in historical perspectives'. In T. Monod (ed.) *Pastoralism in Tropical Africa*. London: Oxford University Press, pp. 406–25.

Jacobsohn, M. 1998 (1990). *Himba: Nomads of Namibia*. Cape Town: Struik.

Johannes, R.E. 1987. 'Primitive myth'. In *Nature* 325: 478.

Johnson, A. 1997. 'Processes for effecting community participation in the establishment of protected areas: a case study of the Crater Mountain Wildlife Management Area'. In C. Filer (ed.) *The Political Economy of Forest Management in Papua New Guinea*. Papua New Guinea and London, National Research Institute and International Institute for Environment and Development, pp. 391–428.

Johnson, C. and Abul T. Hawa, 1999. *Local Participation in Jordanian Protected Areas: Learning from our Mistakes*. Paper given at the Conference, 'Displacement, Forced Settlement, and Conservation', St. Anne's College, 9–11 September.

Jones, B.T.B. 1999. 'Community-based natural resource management in Botswana and Namibia: An inventory and preliminary analysis of progress'. *Evaluating Eden Series Discussion Paper* 6, London: IIED.

Kidder, and W. Balée, 1998. 'Epilogue'. In W. Balée (ed.) *Advances in Historical Ecology*, New York: Columbia, pp. 405–10.

Kirsch, S. 2001. 'Lost worlds: Environmental disaster, "culture loss" and the law', *Current Anthropology*, 42: 2, pp. 167–98.

Koester, D. 2000. *When the Fat Raven Sings: Mimesis and Environmental Alterity in Kamchatka's Environmental Age*. Paper presented at the conference 'Postsocialisms in the Russian North'. Max Planck Institute for Social Anthropology, Halle (Saale) Germany.

Koliesnik, A., G. Isaev, and B. Pietrov. 1995. 'Ecology in the mirror of sociology'. In *Scientific Tatarstan* 3: 40–8.

Koppert, G. J. A., Dounias, A., Froment and P. Pasquet, 1993. 'Food consumption in three forest populations of the southern coastal area of Cameroon: Yassa – Mvae – Bakola'. In C.M. Hladik, A. Hladik, O.F. Linares, H. Pagezy, A. Semple, and M. Hadley (eds) *Tropical Forests, People and Food: Biocultural Interactions and Applications to Development*. Man and the Biosphere series, Volume 13. UNESCO: Paris.

Krech, S. 1999. *The Ecological Indian: Myth and History*. New York: W.W. Norton & Company.

Kuehls, T. 1996. *Beyond Sovereign Territory: The Space of Ecopolitics*. Minneapolis: University of Minnesota Press.

Kuklick, H. 1991. *The Savage Within: The Social History of British Anthropology*. Cambridge: Cambridge University Press.

Kummer, D.M. 1992. *Deforestation in the Postwar Philippines*. Manila: Ateneo University Press.

La Gaceta – Diario Oficial, 18.06.1999.

Lancaster, W. 1981. *The Rwala Bedouin Today*. Cambridge: Cambridge University Press.

Langowiski V.B.R. n.d. (195?) Contribuição para o estudo dos usos e costumes do praieiro do litoral de Paranaguá. *Cadernos do Museu de Arqueologia e Artes Populares*: 77–101.

Lathrap, D.W. 1977. 'Our father the cayman, our mother the gourd: Spinden revisited, or a unitary model for the emergence of agriculture in the New World'. In C.A. Reed (ed.) *Origins of Agriculture*. The Hague: Mouton, pp. 713–51.

Latour, B. 1993. *We Have Never Been Modern*. Cambridge, MA: Harvard University Press.

Latour, B. 1998. 'To modernise or ecologise? That is the question' (trans. C. Cussins). In B. Braun, and N. Castree (eds) *Remaking Reality: Nature at the Millenium*: London and New York: Routledge, pp. 221–42.

La Tribuna. Con apoyo de proyecto Sí-a-Paz: Doscientas familias abandonan la reserva biológica Indio Maíz. 16.04.1998.

Lau, B. 1987. *Southern and Central Namibia in Jonker Afrikaner's Time*. Windhoek.

Leach, M. and R. Mearns (eds) 1996. *The Lie of the Land: Challenging Received Wisdom on the African Environment*. London, Oxford and Portsmouth: The International African Institute, James Currey and Heinemann.

Leonardi, V. 1999. *Os Historiadores e os Rios. Natureza e Ruína na Amazônia Brasileira*. Brasília: Paralelo 15/Universidade de Brasília.

Lévi-Strauss, C. 1971 [1955]. *Tristes Tropiques*. New York: Atheneum.

Lévi-Strauss, C. 1973–81. *Mythologiques* I–IV. London: Cape.

Lewis, H.T. 1989. 'A parable of fire: Hunter-gatherers in Canada and Australia'. In R.E. Johannes (ed.) *Traditional Ecological Knowledge: A Collection of Essays*. Cambridge: IUCN, pp. 11–19.

Lewis, H. T. 1992. 'The technology and ecology of nature's custodians: anthropological perspectives on Aborigines and national parks'. In J. Birckhead, T. De Lacy, and L. Smith (eds) *Aboriginal Involvement in Parks and Protected Areas*. Canberra: Aboriginal Studies Press, pp. 15–28.

Li, T.M. 1999. 'Marginality, power and production: Analysing upland transformations'. In T.M. Li (ed.) *Transforming the Indonesian Uplands*. Amsterdam: Harwood.

Lima, T.S. 1999. 'The two and its many: Reflections on perspectivism in a Tupi cosmology'. *Ethnos* 64(1): 107–31.

Lindsay, W. 1987. 'Integrating parks and pastoralists: Some lessons from Amboseli'. In D. Anderson and R. Grove (eds) *Conservation in Africa: People, Policies and Practice*. Cambridge: Cambridge University Press, pp. 150–67.

Luchiari, M.T.D.P. n.d. A relação do homem com o meio ambiente no universo caiçara. Unpublished paper.

Luke, T. W. 1995. 'On environmentality: Geo-power and eco-knowledge in the discourses of contemporary environmentalism'. *Cultural Critique*, 31: 57–81.

Luke, T. W. 1997. *Ecocritique: Contesting the Politics of Nature, Economy, and Culture*. Minneapolis: University of Minnesota Press.

Mabey, N., S. Hall, C. Smith, and S. Gupta, 1997. *Argument in the Greenhouse: The International Economics of Global Warming*. London: Routledge.

MacKenzie, J.M. 1987. 'Chivalry, social darwinism and ritualised killing: the hunting ethos in Central Africa up to 1914'. In D. Anderson and R. Grove (eds) *Conservation in Africa: People, Policies and Practice*. Cambridge: Cambridge University Press, pp. 41–61.

MacKenzie, J. M. 1988. *The Empire of Nature: Hunting, Conservation and British Imperialism*, Manchester, Manchester University Press.

Maldidier, C. and T. Antillón, 1996. 'Deforestación y frontera agrícola en Nicaragua'. In *Frontera agrícola en Nicaragua*. Managua: UNAM, pp. 5–24.

Malinowski, B. 1935. *Coral Gardens and their Magic: A Study of the Methods of Tilling the Soil and of Agricultural Rites in the Trobriand Islands* Vol. 1–2. London: George Allen & Unwin.

Manning, R. 1989. 'The nature of American visions and revisions of wilderness'. *Natural Resource Journal*, 29: 25–40.

Mantovani W. 1990. A dinâmica das florestas na encosta atlântica. *Simpósio sobre Ecossistemas da Costa Sul e Sudeste Brasileira Estrutura, Função e Manejo*. Águas de Lindóia: Academia de Ciências do Estado de São Paulo, pp. 304–13.

Mantovani W. 1993. *Estrutura e dinâmica da Floresta Atlântica na Juréia, Iguape-SP*. Post-Doc. thesis, Instituto de Biociências da Universidade de São Paulo.

Marcílio, M.L. 1986. *Caiçara terra e população. Estudo de demografia histórica e da história social de Ubatuba*. São Paulo: Edições Paulinas/CEDHAL.

Martin, C. 1978. *Keepers of the Game: Indian–animal Relationships and the Fur Trade*. Berkeley: University of California Press.

Masri, A. 1991. *The Tradition of Hema as a Land Tenure Institution in Arid Land Management: The Syrian Arab Republic*. Rome: FAO.

Maybury-Lewis, D. (ed.) 1979. *Dialectical Societies*. Cambridge, MA: Harvard University Press.

McCallum, R. and N. Sekhran. 1997. *Race for the Rainforest: Evaluating Lessons from an Integrated Conservation and Development 'Experiment' in New Ireland, Papua New Guinea*. Papua New Guinea Biodiversity Conservation and Resource Management Programme, Department of Environment and Conservation (PNG)/United Nations Development Programme – Global Environmental Facility (GEF).

McCracken, J. 1987. 'Conservation priorities and local communities'. In D. Anderson and R. Grove (eds) *Conservation in Africa: People, Policies and Practice*. Cambridge: Cambridge University Press, pp. 63–78.

McCully, P. 1996. *Silenced Rivers: The Ecology and Politics of Large Dams*. London: Zed Books.

McKinnon, S. 1991. *From a Shattered Sun: Hierarchy, Gender and Alliance in the Tanimbar Islands*. Madison: University of Wisconsin Press.

Meadows, D.H., D.L. Meadows, J. Randers, and W.W. Bahrens, 1992. *The Limits to Growth: A Report from the Club of Rome's Project on the Predicament of Mankind*, (2nd edition). London: Earthscan.

Meggers, B.J. 1971. *Man and Culture in a Counterfeit Paradise*. Chicago: Aldine.

MET (Ministry of Environment and Tourism) 1995a. *Policy Document: Promotion of community-based tourism*. Windhoek.

MET (Ministry of Environment and Tourism) 1995b. *Policy Document: Wildlife management, utilisation and tourism in communal areas*. Windhoek.

MET (Ministry of Environment and Tourism) 1995c. *Questions and answers about communal area conservancies*. Ministry of Environment and Tourism, Windhoek.

Milanelo, M. 1992. 'Comunidades tradicionais do Parque Estadual da Ilha do Cardoso e a ameaça do turismo emergente'. In *Congresso Nacional sobre Essências Nativas*. São Paulo: Instituto Florestal – SP. pp. 1109–17.

Miller, D. 1987. *Material Culture and Mass Consumption*. Cambridge, MA: Basil Blackwell.

Miller, D. 1994. *Modernity: An Ethnographic Approach: Dualism and Mass Consumption in Trinidad*, Oxford and New York, Berg.

Miller, D. (ed.) 1995a. *Acknowledging Consumption: A Review of New Studies*, London and New York, Routledge.

Miller, D. 1995b. 'Introduction: Anthropology, modernity and consumption'. In D. Miller (ed.) *Worlds Apart: Modernity through the Prism of the Local*. London and New York, Routledge, pp. 1–22.

Miller, D. 1995c. Consumption and Commodities, *Annual Review of Anthropology* 24: 141–61.

Milton, K. 1993. 'Environmentalism and anthropology'. In K. Milton (ed.) *Environmentalism: The View from Anthropology*. London: Routledge, pp. 1–17.

Milton, K. 1996. *Environmentalism and Cultural Theory: Exploring the Role of Anthropology in the Environmental Discourse*. London: Routledge.

Mitschein, T., Miranda, H., and Paraense, M. 1989. *Urbanização Selvagem e Proletarização Passiva* / Savage Urbanization and Passive Proletarianization. Belem: CEJUP–NAEA.

Moore, D.S. 1996. 'Marxism, culture, and political ecology: Environmental struggles in Zimbabwe's Eastern Highlands'. In R. Peet and M. Watts (eds) *Liberation Ecologies: environment, development, social movements*. London: Routledge, pp. 46–68.

Moran, E. 1981. Developing the Amazon. Bloomington: University of Indiana Press.

Moran, E. 1993. *Through Amazonian Eyes: The Human Ecology of Amazonian Populations*. Iowa City: University of Iowa Press.

Moran, E. (ed.) 1983. *The Dilemmas of Amazonian Development*. Boulder: Westview Press.

Morrison, J. 1993. *Protected Areas and Aboriginal Interests in Canada*. Toronto: WWF Canada Discussion Paper.

Morris-Suzuki, T. 2000. 'For and against NGOs: The politics of the lived world'. *New Left Review*. March-April: 63–84.

Moseley, C., and R. E. Asher, 1994. *Atlas of the World's Languages*. London: Routledge.

Mourão, F.A.A. 1971. *Os pescadores do litoral sul de São Paulo. Um estudo de sociologia diferencial*. Ph.D. thesis, Faculdade de Filosofia Letras e Ciências Humanas da Universidade de São Paulo.

Mouton, D., T. Mufeti, and H. Kisting, 1997. 'A preliminary assessment of the land cover and biomass variations in the Huab'. In D. Ward (ed.) *Land Degradation in the pro-Namib*. Windhoek, Unpublished Report of the Summer Desertification Project (DRFN) 1996 funded by United States Agency for International Development and Swedish International Development Agency.

Murphree, M.W. 1999. Theme presentation on 'governance and community capacity'. Presentation at CASS/PLAAS Inaugural Meeting on Community Based Natural Resource Management in Southern Africa, Harare, September.

Murphy, Y. and R. Murphy, 1974. *Women of the Forest*. New York: Columbia University.

Mussolini, G. 1980. *Ensaios de antropologia indígena e caiçara*. Rio de Janeiro: Paz e Terra.

Nabhan, G.P., D. House, S.A. Humberto, W. Hodgson, H.S. Luis, and M. Guadalupe, 1991. 'Conservation and use of rare plants by traditional cultures of the US/Mexico borderlands'. In Oldfield, M. and Alcorn, J. (eds), *Biodiversity: Culture, Conservation and Ecodevelopment*. Boulder: Westview, pp. 127–46.

Naess, A. 1989. *Ecology, Community and Lifestyle*. Cambridge: Cambridge University Press.

Nash, R. 1982 [1967]. *Wilderness and the American Mind*. New Haven and London: Yale University Press.

National Planning Commission (NPC) 1991. *Population and Housing Census, Preliminary Report*. Windhoek.

Nativenet, http://nativenet.uthscsa.edu/archive/nl/9209/0053.html (accessed June 2000).

Nederveen Pieterse, J. 1992. 'Emancipations, modern and postmodern'. *Development and Change*, 23(3): 5–41.

Neumann, R.P. 1997. 'Primitive ideas: Protected area buffer zones and the politics of land in Africa'. *Development and Change*, 28(4): 559–82.

Neumann, R.P. 1998. *Imposing Wilderness: Struggles over Livelihood and Nature Preservation in Africa*. Berkeley: University of California Press.

Noffs, P.S. 1988. *Os Caiçaras de Toque-Toque Pequeno. Um estudo de mudança espacial*. Ma. diss., Faculdade de Filosofia Letras e Ciências Humanas da Universidade de São Paulo.

Nogueira-Neto, P. 1991. Juréia. In *Estações Ecológicas uma saga de ecologia e de política ambiental*. São Paulo: Banespa, pp. 41–5.

Norton, B.G. 1991. *Toward Unity among Environmentalists*. Oxford: Oxford University Press.

Norton-Griffiths, M. 1996. 'Property rights and the marginal wildebeest: An economic analysis of wildlife conservation options in Kenya'. *Biodiversity and Conservation*, (5): 1557–77.

Norton-Griffiths, M. and C. Southby, 1995. 'The opportunity costs of biodiversity conservation in Kenya'. *Ecological Economics*, (12): 125–39.

Novellino, D. 1997. *Social Capital in Theory and Practice: The Case of Palawan*. Rome: FAO.

Novellino, D. 1998. 'Sacrificing people for the trees': the cultural cost of forest conservation on Palawan Island (Philippines). *Indigenous Affairs*, 4: 4–14.

Novellino, D. 1999a. 'The ominous switch: from indigenous forest management to conservation – the case of the Batak on Palawan Island, Philippines'. In M. Colchester, and C. Erni (eds) *Indigenous Peoples and Protected Areas in South and Southeast Asia*, Document No. 97, Copenhagen: IWGIA, pp. 250-97.

Novellino, D. 1999b. 'An account of agency in some Pälawan attitudes toward illness and healing practices'. In A. Guerci (ed.) *Meetings Between Medicines*. Genova: Erga Edizioni, pp. 298-311.

Novellino, D. 1999c. 'Prohibited food and dietary habits among the Batak of Palawan Island (Philippines)'. In A. Guerci (ed.) *Cultural Food. From Food to Culture. From Culture to Food*. Genova: Erga edizioni, pp. 145-54.

Novellino, D. 1999d. 'The Emergency Role of Palm Food in Palawan Island (Philippines)'. In A. Guerci, (ed.) *Food and Body. From Food to Culture. From Culture to Food*. Genova: Erga Edizioni, pp. 52-74.

Novellino, D. 1999e. 'Towards an understanding of Pälaqwan rock drawings: Between visual and verbal expression'. *Rock Art Research*, 16(1): 3–24.

Novellino, D. 2000a. 'Forest conservation in Palawan'. *Philippine Studies*, 48: 347-72.

Novellino, D. 2000b. 'Recognition of ancestral domain claims on Palawan island, the Philippines: Is there a future?' *Land reform: land settlement and cooperatives 2000/1*. Rome: FAO.

Nugent, S. 1993. *Amazonian Caboclo Society: An Essay on Invisibility and Peasant Economy*. Oxford: Berg.

Nugent, S. 1997. 'The coordinates of identity in Amazonia: At play in the fields of culture'. *Critique of Anthropology*, 17(1): 33–51.

Nugent, S. 2000. 'Good risk, bad risk: Reflexive modernisation and Amazonia'. In P. Caplan (ed.) *Risk Revisited*. London: Pluto Press, pp. 226–48.

Nygren, A. 1999. 'Local knowledge in the environment-development discourse: from dichotomies to situated knowledges'. *Critique of Anthropology*, 19 (3): 267–88.

Nygren, A. 2000. 'Environmental narratives on protection and production: Nature-based conflicts in Río San Juan, Nicaragua'. *Development and Change*, 31 (4): 807–830.

Odendaal Report. 1964. *Report of the Commission of Enquiry into South West African affairs 1962–1963*. Pretoria.

Offen, K. 1992. *Productos forestales no maderables y su manejo campesino en la zona de amortiguamiento, Sí-A-Paz*. Managua (mimeogr).

Oldfield, M. and J. Alcorn (eds) 1991. *Biodiversity: Culture, Conservation and Ecodevelopment*. Boulder: Westview Press.

Oliveira, R.R. 1999. *O rastro do homem na floresta sustentabilidade e funcionalidade da mata atlântica sob manejo caiçara*. Ph.D. thesis, Programa de Pós Graduação em Geografia da Universidade Federal do Rio de Janeiro.

O'Riordan, T. 1981. *Environmentalism*. London: Pion Limited.

Ó Tuathail, G and S. Dalby, 1998. Introduction: Rethinking geopolitics – towards a critical geopolitics. In G. Ó Tuathail and S. Dalby (eds) *Rethinking Geopolitics*. London and New York: Routledge, pp. 1–15.

Overing, J. 1981. 'Amazonian anthropology'. *Journal of Latin American Studies*, 13(1): 151–64.

Pasquet, P., A. Froment, and R. Ohtsuka, 1993. 'Adaptive aspects of food consumption and energy expenditure – background'. In M.C. Hladik, A. Hladik, O. F. Linares, H. Pagezy, A. Semple, and M. Hadley (eds) *Tropical Forests, People and Food: Biocultural Interactions and Applications to Development*. Man and the Biosphere series, Volume 13. UNESCO: Paris, pp. 249-56.

Pauwels, S., D.D. Coppet, and R.J. Parkin 1985. 'Some important implications of marriage alliance: Tanimbar, Indonesia'. In Barnes, R.H. et al. (eds) *Contexts and Levels: Anthropological Essays on Hierarchy*. Oxford: JASO, pp. 131-38.

Pearce, D. W., and R. K. Turner. 1990. *Economics of Natural Resources and the Environment*. Hemel Hempstead: Harvester Wheatsheaf.

Pearl, M. C. 1994. 'Local initiatives and the rewards for biodiversity conservation: Crater Mountain Wildlife Management Area, Papua New Guinea.' In D. Western and S. Strum (eds) *Natural Connections: perspectives in community-based conservation.* Washington DC: Island Press, pp. 193–214.

Peet, R. and M. Watts. 1996. 'Liberation ecology: development, sustainability, and environment in an age of market triumphalism'. In R. Peet and M. Watts (eds) *Liberation Ecologies: Environment, Development, Social Movements.* London: Routledge, pp. 1–45.

Peluso, N.L. 1992. *Rich Forests, Poor People: Resource Control and Resistance in Java.* Berkeley: University of California Press.

Pererjolkin, L. and Y. Figatner. 1997. 'Environmental movements in Moscow'. In K.L. Pickvance, N. Manning, and C. Pickvance (eds) *Environmental and Housing Movements: Grassroots Experience in Hungary, Russia and Estonia.* Avebury: Aldershot, pp. 198–255.

Perrett, R.W. 1998. 'Indigenous rights and environmental justice'. *Environmental Ethics,* 20: 377–91.

Pepper, D. 1986. *The Roots of Modern Environmentalism.* London and New York: Routledge.

Peters, C., A. Gentry, and R. Mendelsohn, 1989. 'Valuation of an Amazonian forest'. *Nature* 339: 655–56.

Pickvance, K. L., N. Manning, and C. Pickvance (eds) 1997. *Environmental and Housing Movements: Grassroots Experience in Hungary, Russia and Estonia.* Avebury: Aldershot.

Pierson, D. and C.B. Teixeira, 1947. Survey de Icapara. *Sociologia* 9.

Pigg, S.L. 1992. 'Inventing social categories through place: Social representations and development in Nepal'. *Comparative Studies in Society and History,* 34(3): 491–513.

Pimbert, M. and J. Pretty, 1995. *Parks, People and Professionals: Putting Participation into Protected Area Management.* Geneva: United Nations Research Institute for Social Development (UNIRSD). Discussion Paper 57.

Polanyi, K. 1957. 'The economy as an instituted process'. In: Karl Polanyi (ed.) *Trade and Market in the Early Empires.* New York: Free Press, pp. 243–70.

Popkov, Iu. V. 1994. *Etnosotsial'nye i pravovye protsessy v Evenkii.* Novosibirsk: Sibirskoe Otdelenie RAN.

Por, F.D. 1992. *Sooretama: The Atlantic Rain Forest of Brazil.* The Hague: SPB Academic Publishing.

Powers, W.K. 1975. *Oglala Religion.* Lincoln, NE: University of Nebraska Press.

Pulido, L. 1996. *Environmentalism and Economic Justice: Two Chicano Struggles in the Southwest.* Tucson, AZ: University of Arizona Press.

Pyne, S.J. 1997. *World Fire: The Culture of Fire on Earth.* Seattle: University of Washington Press.

Rabella, J. 1995. *Aproximación à la historia de Río San Juan (1500–1995).* Managua: Solidaridad Internacional.

Rae, J. 1999. *Tribe and State: Management of the Syrian Steppe.* PhD thesis, University of Oxford.

Rae, J., G. Arab, and T. Nordblom, 1999 *Rejection of Custom: Aridland Conservation in Syria.* Paper presented at Conference on Displacement, Forced Settlement and Conservation. 9–11 September 1999, Queen Elizabeth House, University of Oxford.

Ramos, A.R. 1998. *Indigenism: Ethnic Politics in Brazil.* Madison: University of Wisconsin Press.

Rangan, H. 1996. 'From Chipko to Uttaranchal: Development, environment and social protest in the Garhwal Himalayas, India'. In R. Peet and M. Watts (eds) *Liberation Ecologies: Environment, Development, Social Movements.* London and New York: Routledge, pp. 205–26.

Rappaport, R. 1968. *Pigs for the Ancestors.* New Haven: Yale University Press.

Republic of the Philippines, Congress of the Philippines. 1992a. Republic Act no. *7586 (National Integrated Protected Areas System Act).* Manila.

Republic of the Philippines, Congress of the Philippines. 1992b. Republic Act no. 7611. Manila.

Research and Conservation Foundation of Papua New Guinea. 1999. *Crater Mountain Wildlife Management Area project: Terms of reference.* Project document.

Research and Conservation Foundation of Papua New Guinea/ Wildlife Conservation Society. 1995. Crater Mountain Wildlife Management Area: A model for testing the linkage of community-based enterprises with conservation of biodiversity (BCN Implementation Grant Proposal), Port Moresby, RCF & WCS.

Ribeiro D. 1995. *O povo brasileiro: a formação e o sentido do Brasil.* Sao Paulo: Compantiia das Letras.

Ribot, J. 1998. 'Theorizing access: Forest profits along Senegal's charcoal commodity chain'. *Development and Change,* 29: 307–41.

Riviére, P. 1993. 'The amerindianization of descent and affinity'. *L'Homme,* 23(2–4): 507–516.

Robbins, P. 2000. 'Pastoralism and community in Rajasthan: interrogating categories of arid land development'. In A. Agarwal, and K. Sivaramakrisnan (eds) *Agrarian Environments: Resources, Representation and Rule in India.* Durham, NC: Duke University Press, pp. 191–215.

Rocheleau, D., Steinberg, P. E. and Benjamin, P. A. 1995. 'Environment, development, crisis, and crusade: Ukambani, Kenya, 1890–1990'. *World Development*, 23(6): 1037–51.

Roe, D., J. Mayers, Grieg-Gran, M., Kothari, A., Fabricius, C. and Hughes, R. 2000. *Evaluating Eden: Exploring the Myths and Realities of Community-Based Wildlife Management*. London: IIED.

Roeder, H., 1996. *Socio-economic study of the Bishri mountains*. Cologne: Deutsche Gesellschaft für Technische Zusammenarbeit (GTZ).

Rohde, R.F. 1993. 'Afternoons in Damaraland: Common land and common sense in one of Namibia's former "homelands".' *Centre of African Studies, Edinburgh University, Occasional Paper* 41.

Roosevelt, A.C. 1991. *Moundbuilders of the Amazon*. New York: Academic Press.

Roosevelt, A.C. (ed.) 1994. *Amazonian Indians: From Prehistory to the Present*. Tucson, AZ: University of Arizona Press.

Rootes, C. and Davis, H. (eds) 1994. *A New Europe? Social Change and Political Transformation*. London: UCL Press.

Rosaldo, R. 1986. 'Ilongots hunting as story and experience'. In V. Turner and Bruner, M. E. (eds) *The Anthropology of Experience*. Chicago: University of Illinois Press, pp. 97–138.

Roseman, M. 1991. *Healing Sounds from the Malaysian Rainforest: Temiar music and medicine*. Los Angeles: University of California Press.

Rougeulle, M.D. 1989. 'Pescas Artesanais de Guaraqueçaba'. In *Encontro de Ciências Sociais e o Mar*. São Paulo: PPCAUB – Universidade de São Paulo, pp. 281–8.

Rychkov, K. M. 1922–1923. 'Eniseiskie Tungusy'. In *Zemlevedenie*, pp. 1–67; 69–106; 107–149.

Sachs, W. (ed.) 1993. *Global Ecology: A New Arena of Political Conflict*. London: Zed Books.

Sales, R.J.R. 1988. 'Aspectos da pesca artesanal na região lagunar de Iguape-Cananéia'. *Encontro de Ciências Sociais e o Mar 2*. São Paulo: *PPCAUB*- Universidade de São Paulo, pp. 63–75.

Sales R.J.R., and A.C.C. Moreira. 1994. *Estudo de viabilidade de implantação de reservas extrativistas no Domínio Mata Atlântica, município de Cananéia*. São Paulo: *NUPAUB*- Universidade de São Paulo.

Saloman, F. and S. Schwartz (eds) 1999. *Introduction: The Cambridge History of the Native Peoples of the Americas. Vol. III, South America*. Cambridge: Cambridge University Press, pp. 1–18.

Sampaio, T. 1987. *O tupi na geografia nacional*. Brasília: Brasiliana.

Sanderson, S.E., and K. Redford 1997. 'Biodiversity politics and the contest for ownership of the world's biota'. In R. Kramer, C. van Schaik, and J. Johnson (eds) *Last Stand: Protected Areas and the Defense of Tropical Biodiversity*. Oxford: Oxford University Press, pp. 115–32.

Sanford, S. 1983. *Management of Pastoral Development in the Third World*. London: John Wiley and Sons.

Schaden, E. 1954. 'Os primitivos habitantes do território paulista'. *Revista de História FFLCH–USP*, 5(18): 385–406.

Schroeder, R.A. 1999. 'Geographies of environmental intervention in Africa'. *Progress in Human Geography*, (23)3: 359–78.

Schwarzc, L. 1999. *The Spectacle of the Races*. New York: Hill and Wang.

Scoones, I. 1999. 'New ecology and the social sciences: what prospects for a fruitful engagement?' *Annual Review of Anthropology*, 26: 479–507.

Scott, C. 1996. 'Science for the West, myth for the rest? The case of James Bay Cree knowledge construction'. In Laura Nader (ed.) *Naked Science: Anthropological Inquiry into Boundaries, Power, and Knowledge*. New York: Routledge, pp. 69–86.

Scott, J. 1998. *Seeing Like a State: How Certain Schemes to Improve the Human Condition Have Failed*. New Haven: Yale University Press.

Seely, M.K., and K.M. Jacobson, 1994. 'Guest editorial. Desertification and Namibia: a perspective'. *Journal of African Zoology* (108): 21–36.

Seely, M.K., Hines, C. and A.C. Marsh, 1995. 'Effects of human activities on the Namibian environment as a factor in drought susceptibility'. In R. Moorsom, J. Franz, and M. Mupotola (eds) *Coping with Aridity: Drought Impacts and Preparedness in Namibia*. Frankfurt: Brandes & Apsel. pp. 51–61.

Segura, O., Kaimowitz, D. and J. Rodríquez, 1997 *Políticas forestales en Centro América: Análisis de las restricciones para el desarrollo del sector forestal*. San Salvador: EDICPSA.

Serna, C.B. 1990. 'Rattan resources supply situation'. In N.K. Toreta and E.H. Belen (eds) *Rattan: Proceedings of the National Symposium/Workshop on Rattan, Cebu City, 1988*.

Setti K. 1985. *Ubatuba nos cantos das praias. Estudo do caiçara paulista e de sua produção musical*. São Paulo: Ática.

Shenon, P. and R. Full, 1969. *An appraisal of the mineral resources in the lands of the Sioux nation acquired by the United States under Treaty of April 26, 1868, ratified February 16, 1869 and the Act of February 28, 1877 – Indian Claims Commission Docket Nos. 74 and 74b – Sioux Tribe, et al., vs. United States of America*. 2 vols. Salt Lake, Utah.

Shivute, O. 1998. 'Conservancy plan sparks tribal row', *The Namibian*, 16 July, Windhoek.

Shoup, J. 1990. 'Middle Eastern sheep pastoralism and the Hima system'. In J. Galaty and D. Johnson (eds) *The World of Pastoralism: Herding Systems in Comparative Perspective*. New York: The Guilford Press, pp. 195–215.

Sillitoe, P. 1996. *A Place Against Time: Land and the Environment in the Papua New Guinea Highlands*. Amsterdam: Harwood Academic Publishers.

Simpson, R.D. and R.A. Sedjo, 1996, 'Paying for the conservation of endangered ecosystems: a comparison of direct and indirect approaches'. *Environment and Development Economics*, (1): 241–57.

Silva, A.C. 1975. *O litoral norte do Estado de São Paulo. Formação de uma região periférica*. Vol. 20. São Paulo: IGEOG – Universidade de São Paulo.

Silva, J.G.S. 1993. *Caiçaras e jangadeiros cultura marítima e modernização no Brasil*. São Paulo: CEMAR – Universidade de São Paulo.

Silva, Y.M.F.A. 1979. *Trindade sobrevivência e expropriação*. MA diss., Pontifícia Universidade Católica de São Paulo.

Silverstone, R., and Hirsch, E. (eds) 1992 *Consuming Technologies: Media and Information in Domestic Spaces*. London: Routledge.

Simão A. and F. Goldman, 1958. *Itanhaém. Estudo sôbre o desenvolvimento econômico e social de uma comunidade litorânea*. São Paulo: FFCL – Universidade de São Paulo.

Singh, S. P. 1998. 'Chronic disturbance, a principal cause of environmental degradation in developing countries'. *Environmental Conservation*, 25(1): 1–2.

Siqueira, P. 1989. 'Os Caiçaras do Litoral Norte do Estado de São Paulo'. In *Encontro de Ciências Sociais e o Mar*. São Paulo: PPCAUB, Universidade de São Paulo / Ford Foundation / IUCN, pp. 263–72.

Sirina, A.A. 1999. *Rodovye Obshchiny Malochislennykh Narodov Severa v Resublike Sakha (Yakutiia): Shag k Samoopredleniiu?* Moskva: Institut Etnografii i Etnologii.

Sixth District Council of Local Governments. 1976/77. *Energy development in the Sixth District*. Unpublished paper.

Slater, C. 1995. 'Amazonia as Edenic narrative'. In W. Cronon (ed.) *Uncommon Ground: Toward Reinventing Nature*. New York: W. W. Norton & Company, pp. 114–131.

Smith, N. 1999. *The Amazon River Forest*. Oxford: Oxford University Press.

Soulé, M. and M. A. Sanjayan. 1998. Conservation targets: do they help? *Science* 279: 2060–61.

Spyer, P. 1996. 'Diversity with a difference: Adat and the New Order in Aru, eastern Indonesia'. *Cultural Anthropology*, 11(1): 25–50.

Steward, J. 1949. *Handbook of South American Indians*. Washington DC: U.S. Government Printing Office.

Stocking, G. W. 1991. 'Maclay, Kubary, Malinowski: archetypes from the dreamtime of anthropology'. In Stocking, G. W. (ed.) *Colonial Situations: Essays in the Contextualization of Ethnographic Knowledge*. History of Anthropology, 7. Madison: University of Wisconsin Press, pp. 9–74.

Strathern, M. 1980. 'No nature, no culture'. In Strathern, M. and MacCormack, C. (eds) *Nature, Culture and Gender*. Cambridge: Cambridge University Press, pp. 174–222.

Strathern, M. 1992. *After Nature: English kinship in the late twentieth century*. Cambridge: Cambridge University Press.

Strathern, M. 1995. 'The nice thing about culture is that everyone has it'. In M. Strathern (ed.) *Shifting Contexts: Transformations in Anthropological Knowledge*. London and New York: Routledge, pp. 151–76.

Strathern, M. 1999. *Property, Substance and Effect: Anthropological Essays on Persons and Things*. London: Athlone Press.

Stuart M., and Sekhran, N. 1996. *Developing externally financed greenhouse gas mitigation projects in Papua New Guinea's forestry sector: A review of concepts, opportunities and links to biodiversity conservation*. Papua New Guinea, Department of Environment and Conservation/ United Nations Development Program (PNG Biodiversity Conservation and Resource Management Program).

Subler, S. and C. Uhl, 1990. 'Japanese agroforestry in Amazonia: a case study in Tomé-Açu, Brazil'. In Anderson, A. (ed.) *Alternatives to deforestation*. New York: Columbia University Press, pp. 152-66.

Sullivan, S. 1996. *The 'communalization' of former commercial farmland: perspectives from Damaraland and implications for land reform*, Windhoek: Social Sciences Division of the Multi-Disciplinary Research Centre, University of Namibia, Research Report No. 25.

Sullivan, S. 1999. 'Folk and formal, local and national: Damara knowledge and community-based conservation in southern Kunene, Namibia'. *Cimbebasia*, (15): 1–28.

Sullivan, S. 2001. 'Difference, identity, and access to official discourses: Haillom, 'Bushmen', and a recent Namibian ethnography. *Anthropos*, (96): 179–92.

Sullivan, S. in press. 'How sustainable is the communalising discourse of "new" conservation? The masking of difference, inequality and aspiration in the fledgling "conservancies" of north-west Namibia'. In
D. Chatty (ed.) *Displacement, Forced Settlement and Conservation.* Oxford: Berghahn Press.

Sullivan, S. in prep. 'The "wild" and the known: implications of identity and memory for CBNRM in a
Former Namibian "homeland"'. In C. Twyman and M. Taylor (eds) *Entitled to a Living: CBNRM in
Southern Africa.* Oxford.

Sullivan, S. and Ganuses, W.S. in prep. *Faces of Damaraland: Life and Landscape in a Former Namibian 'Homeland'* (Working Title). Windhoek.

Sunderlin, W.D. 1999. *The effects of economic crisis and political change on Indonesia's forest sector,
1997–99.* CIFOR Background Paper, November 13.

Sunderlin, W.D. and Resosudarmo, I.A.P. 1996. *Rates and causes of deforestation in Indonesia: towards
a resolution of the ambiguities.* CIFOR Occasional Paper No. 9

Sunderlin, W.D., I.A.P. Resosudarmo, E. Rianto, and A. Angelsen, 2000 *The effect of Indonesia's economic
crisis on small farmers and natural forest cover in the Outer Islands.* CIFOR Occasional Paper No. 28.

Sutherland, J. 1998. 'Top Award for Namibia'. *The Namibian,* 28 September, Windhoek.

Syrian Arab Republic. 1996. *Steppe Directorate 1996.* Damascus.

Takacs, D. 1996. *The Idea of Biodiversity: Philosophies of Paradise.* Baltimore: John Hopkins University
Press.

Taylor, M. 1999. 'You cannot put a tie on a buffalo and say that is development': Differing priorities in
community conservation, Botswana. Paper Presented at Conference on 'African Environments – Past
and Present', St. Anthony's College, University of Oxford, 5–8 July, 1999.

Thoma, A. and J. Piek, 1997. *Customary Law and Traditional Authority of the San.* Windhoek, Centre for
Applied Social Sciences Paper No. 36.

Thomas, K. 1983. *Man and the Natural World: Changing Attitudes in England, 1500 to 1800.* London:
Allen Lane.

Trigger, D. S. 1998. 'Citizenship and indigenous responses to mining in the Gulf country'. In N. Peterson,
and W. Sanders (eds) *Citizenship and Indigenous Australians: Changing Conceptions and Possibilities.*
Cambridge: Cambridge University Press, pp. 154–66.

Tsing, A.L. 1993. *In the Realm of the Diamond Queen: Marginality in an Out-of-the-Way Place,* Princeton: Princeton University Press

Tuzin, D. 1992. 'Sago subsistence and symbolism among the Ilahita Arapes'h. *Ethnology,* 31: 103–14.

Twyman, C. in press. 'Livelihood opportunity and diversity in Kalahari Wildlife Management Areas,
Botswana: Rethinking community resource management'. *Journal of Southern African Studies.*

United Church of Christ, Commission for Racial Justice. 1987. *Toxic Waste and Race in the United States:
National report on the racial and socio-economic characteristics of communities with hazardous waste
sites.* New York: Public Access Data.

United Nations World Council of Indigenous Peoples. 1994. United Nations Draft Declaration on the
Rights of Indigenous Peoples. Geneva, Switzerland: F/CN.4/Sub.2/ 1994/2/Add.1.

United States Department of the Interior. 1971. *North Central power study.* Washington, DC: U.S. Government Printing Office.

Urry, J. 1995. *Consuming Places.* London: Routledge.

Utting, P. 1993. *Trees, People and Power: Social Dimensions of Deforestation and Forest Protection in
Central America.* London: Earthscan.

Valença, M.M. 1999. *Patron-Client Relations and Politics in Brazil.* London: Research Papers in Environmental & Spatial Analysis 58: 1 –42.

Vasil'ev, V. N. 1908. 'Kratkii ocherk inorodtsev Turkhanskogo Kraia'. *Ezhegodnik Russkogo Antropo-
logicheskogo Obshestva.* Sankt Peterburg: Sankt Peterburzhskii universitet, pp. 57–87.

Vayda, A.P. 1996. *Methods and Explanations in the Study of Human Actions and their Environmental
Effects.* Jakarta: CIFOR/WWF.

Vegacruz, C. 1995. *Inventario de programas y proyectos existentes.* Managua: MARENA (mimeogr.).

Vianna L.P. 1996. *Considerações críticas sobre a construção da idéia de população tradicional no con-
texto das unidades de conservação.* MA. diss., Departamento de Antropologia da Universidade de São
Paulo.

Vianna L.P., C. Adams, and A.C.S. Diegues, 1995. *Conflitos entre populações humanas e áreas naturais
protegidas na Mata Atlântica.* São Paulo: NUPAUB – Universidade de São Paulo.

Viola E. J. and H. R. Leis. 1995. 'A Evolução das Políticas Ambientais no Brasil, 1971–1991 do bisseto-
rialismo preservacionista para o multissetorialismo orientado para o desenvolvimento sustentado'. In D.
J. Hogan, and P. F. Vieira (eds) *Dilemas Socioambientais e Desenvolvimento Sustentável.* Campinas:
Universidade de Campinas, pp. 73–102.

Vitae, Civilis 1995. *Direito de uso de recursos naturais e de propriedade intelectual o caso Juréia*. São Paulo: Vitae Civilis.

Viveiros de Castro, E. 1996. 'Images of nature and society in Amazonian ethnology'. *Annual Review of Anthropology*, 25: 179–200.

Viveiros de Castro, E. 1998. 'Cosmological deixis and Amerindian perspectivism'. *Journal of the Royal Aanthropological Institute* (N.S.) 4: 469–488.

Vivian, J. 1994. 'NGOs and sustainable development in Zimbabwe: No magic bullets'. *Development and Change*, 25(1): 167–193.

Wagner, R. 1979. 'The talk of Koriki: A Daribi contact cult'. *Social Research*, 46: 140–65.

Walker, J.R. 1917. 'The Sun Dance and other ceremonies of the Oglala division of the Teton Dakota'. In *Anthropological Papers* 16(2): 51–221. New York: American Museum of Natural History.

Walker, J.R. 1980. *Lakota Belief and Ritual*. R.J. DeMallie and E.A. Jahner (eds) Lincoln, NE: University of Nebraska Press.

Warren, C. 1993. *Adat and Dinas: Balinese Communities in the Indonesian State*. Oxford: Oxford University Press.

Weaver, J. 1996. 'Introduction: Notes from a miner's canary'. In *Defending Mother Earth: Native American Perspectives on Environmental Justice*. Jace Weaver, (ed.) Maryknoll, NY: Orbis Books, pp. 1–28.

Weiner, D. R. 1988. *Models of Nature: Ecology, Conservation, and Cultural Revolution in Soviet Russia*. Indiana-Michigan Series in Russian and East European Studies. Bloomington: Indiana University Press.

Weiner, D.R. 1999. *A Little Corner of Freedom: Russian Nature Protection from Stalin to Gorbachev*. Berkeley, California: University of California Press.

Wells, M., K. Brandon, and L. Hannah, 1992. *People and Parks: Linking Protected Area Management with Local Communities*. Washington DC: World Bank.

Wenz, P. S. 1988. *Environmental Justice*. Albany, NY: State University of New York Press.

Wenzel, G. 1991. *Animal Rights, Human Rights: Ecology, Economy, and Ideology in the Canadian Arctic*. Toronto: University of Toronto Press.

Werner, W. 1998. *No One Will Become Rich: Economy and Society in the Herero Reserves in Namibia, 1915–1946*. Basel, P. Schlettwein Publishing.

Western, D. and S. Strum (eds) 1994. *Natural Connections: Perspectives in Community-based Conservation*. Washington DC: Island Press.

White Face, C. 1998. 'The seventh generation'. *Rapid City Journal*. 13 May.

Whitehead, N.L. 1993a. 'Ethnic transformations and historical discontinuity in native Amazonia and Guiana, 1500–1900'. *L'Homme*, 126–8: 285–305.

Whitehead, N.L. 1993b. 'Recent research on the native history of Amazonia and Guiana'. *L'Homme*, 126–28: 495–506.

Willems E., Buzios Island. 1966. *A Caiçara community in Southern Brazil* Vol. 20. Seattle and London: University of Washington Press.

Wilmsen, E.N. 1996. 'Introduction: Premises of power in ethnic politics'. In E. Wilmsen, and P. McAllister (eds) *The Politics of Difference: Ethnic Premises in a World of Power*. Chicago: University of Chicago Press, pp. 1–23.

Wilson, L. 1999. *Report of Tura store survey: prices of fish and meat products June 01 to September 20, 1999*. Unpublished manuscript. University of Alberta October 1999.

Winther J., Rodrigues, E. P. F. and Maricondi, M. I. 1989. *Laudo de ocupação da comunidade de São Paulo Bagre, Cananéia (SP)*. São Paulo: PPCAUB – Universidade de São Paulo / Ford Foundation / IUCN.

Wolf, E.R. 1997. Preface and introduction. In *Europe and the People Without History*. London: University of California Press, pp. 1–23.

Wolters, S. (ed.) 1994. *Proceedings of Namibia's National Workshop to Combat Desertification*. Windhoek, NAPCOD.

Women of All Red Nations. 1980. *Pine Ridge Reservation health study: Mass contamination and its direct link with human survival*. Privately printed. Porcupine, SD.

Worster, D. 1977. *Nature's Economy: A History of Ecological Ideas*. Cambridge: Cambridge University Press.

Wrangham, R. 2002. 'Changing Policy Discourses and Traditional Communities, 1960–1999'. In C.J.P. Colfer, and I.A.P. Resosudarmo (eds) *Which Way Forward? People, Forests, and Policymaking in Indonesia*. Washington DC: Resources for the Future, pp. 20–35.

Yanizkii, O.N. 1987. *Ekologicheskaia perspektiva goroda*. Moskva: Mysl.

Ziegler, C.E. 1987. *Environmental Policy in the USSR*. London: Frances Printer Publishers.

Ziker, J.P. 1998. 'Kinship and exchange among the Dolgan and Nganasan of northern Siberia'. *Research in Economic Anthropology*, 19: 191–238.

INDEX

Indigenous Peoples Rights Act
(IPRA), 188n. 5
National Integrated Protected Areas
System (NIPAS), 179–81
Strategic Environmental Plan (SEP),
179–80, 181–83
Pine Ridge Reservation, 110–11, 114,
118n. 14, 118n. 16
Pio-Tura region, 124–33
Crater Mountain project, 127–28, 130
Power Supply Entities of the North Central
and Rocky Mountain Region, 112
protected areas, 94, 123. *See also* Brazil
and Nicaragua
forced removal of inhabitants, 78–79,
94–95
Papua New Guinea, 127–28
Philippines, 179–84
Siberia, 169
Syria, 94, 95–97
Pyne, Stephen, 9

R
Rappaport, Roy, 193
Research and Conservation Foundation
(RCF), 127, 132
Río San Juan, 34–43
Rubber Tappers' Union, 192, 193–4, 201
rural cultures
'authenticity' of, 13
Rural People's Institute for Social Empow-
erment (RISE), 84
Russian Federation, 139–53. *See also*
Evenki Autonomous District, Guard
of Nature, and Soviet Union (for-
mer)
ecological movements, 139–40,
142–49, 150–53
State Committee of the North, 156

S
Sarney, Jose, 196
Saudi Arabia, 99n. 10
Sesfontein, 73–75, 77, 81–82

Siberia, 9–10, 155–70. *See also* Evenki
Autonomous District
Sierra Club, 107, 113
SKEPHI, 60
slavery
Brazil, 22, 30n. 6, 195
social justice and environmental conservation, 102–4, 117, 192, 193, 204
Soviet Union (former). 139–43, 149–50,
151–52, 153n. 3. *See also* Evenki
Autonomous District, Guard of
Nature, and Russia
censorship, 141
ecological movements, 140–42
environmental degradation, 140,
141–42
natural reserves, 8–9
Spruce, Richard, 190
Stocking, George, 2
Survival International, 61
Syria, 87–99. *See also* Badia and Bedouin
colonisation of, 87–89
control of Bedouin, 87–90, 92, 93, 95
international development, 90–92

T
Taimyr Autonomous District, 160–61
Tanimbar Islands. *See also* forests and
Inhutani
agriculture, 54–55, 620
colonisation of, 53, 63
description of, 52–55
and Asian economic crisis, 55
forestry, 55–64
local government, 53–54, 64, 65
protests against logging, 52, 59–61, 64,
65
relations with the state, 53–54, 62–63
resource access, 54–56, 57, 61–64
Tanimbarese Intellectuals Association
(*Ikatan Cendekiawan Tanimbar
Indonesia* – ICTI), 52–53, 57,
59–61, 64–65, 66n. 9